THE NEW CLARENDON BIBLE
OLD TESTAMENT
VOLUME I

—

THE HISTORY AND RELIGION
OF ISRAEL

The History and Religion of Israel

BY

G. W. ANDERSON, M.A., D.D., F.B.A.

PROFESSOR OF HEBREW
AND OLD TESTAMENT STUDIES
UNIVERSITY OF EDINBURGH

OXFORD UNIVERSITY PRESS

Oxford University Press, Walton Street, Oxford OX2 6DP

OXFORD LONDON GLASGOW
NEW YORK TORONTO MELBOURNE WELLINGTON
KUALA LUMPUR SINGAPORE HONG KONG TOKYO
DELHI BOMBAY CALCUTTA MADRAS KARACHI
NAIROBI DAR ES SALAAM CAPE TOWN

First published 1966
Reprinted 1966 (with corrections)
1969, 1971, 1974, 1975, 1978, 1979

Printed in Great Britain
at the University Press, Oxford
by Eric Buckley
Printer to the University

PUBLISHER'S PREFACE

WHEN it became necessary, a year or two ago, to contemplate revision of the Old Testament volumes of the Clarendon Bible series, the publishers were faced with two important decisions: first, on what text the revision should be based, and second, whether any significant change should be made in the form and plan of the series.

It seemed to them, after taking the best available advice, that the Revised Version could not expect to hold the field for much longer in face of the developments in scholarship which have taken place since its publication in the eighteen eighties, and which have been reflected in more recently published versions. On the other hand, the New English Bible Old Testament is not yet published, and even after its publication it will be some little time before its usefulness for schools and universities can be evaluated. In these circumstances, the Revised Standard Version has seemed the obvious choice, the more particularly because of the recent decision by the Roman Catholic hierarchy to authorize the use of their own slightly modified version in British schools. The publishers would like to express their gratitude to the National Council of Churches of Christ in the United States of America for the permission, so readily given, to make use of the R.S.V. in this way.

With regard to the form of the series, the success of the old Clarendon Bible over the years has encouraged them to think that no radical change is necessary. As before, therefore, subjects requiring comprehensive treatment are dealt with in essays, whether forming part of the introduction or interspersed among the notes. The notes themselves are mainly concerned with the subject-matter of the books and the points of interest (historical, doctrinal, etc.) therein presented; they deal with the elucidation of words, allusions, and the like only so far as seems necessary to a proper comprehension of the author's meaning. There will, however, be some variations in

the content and limits of each individual volume, and in particular it is intended that a fuller treatment should be given to
Genesis, and to the Psalms.

The plan is to replace the volumes of the old series gradually
over the next few years, as stocks become exhausted.

AUTHOR'S PREFACE

THIS volume, the first of the Old Testament series in the New Clarendon Bible, differs in two important ways from the corresponding volume contributed by Dr. W. L. Wardle to the original Clarendon Bible. The history and the religion are treated not separately, but together, so far as may be, in chronological sequence; and the historical narrative, which in Dr. Wardle's book ended with Ezra, is here continued as far as the Maccabaean Revolt. The reasons for the first of these changes are indicated in the introductory chapter. The second seemed desirable in order to cover, however briefly, the entire period to which the canonical Scriptures of the Old Testament refer, and, more particularly, to provide the necessary historical background for the book of Daniel and the rise of apocalyptic.

In accordance with editorial policy, quotations from the text of Scripture follow the Revised Standard Version, of which incomparably the most useful edition is *The Oxford Annotated Bible* (edited by Herbert G. May and Bruce M. Metzger). To save space, verbatim quotations have not been given from extra-biblical texts which illustrate the Old Testament narrative; but references have been given to the two useful collections: *Documents from Old Testament Times* (edited by D. Winton Thomas) (*DOTT*) and *The Ancient Near East* (edited by James B. Pritchard) (*ANE*). On questions of biblical geography the reader is recommended to consult *The Oxford Bible Atlas* (edited by Herbert G. May).

I wish to express my indebtedness to the authorities of the Clarendon Press, for their courtesy, patience, and helpfulness, to Mrs. E. S. K. Paterson and my wife, who prepared the final typescript, and to the Revd. Martin Reid, who prepared the indexes.

G. W. ANDERSON

November 1965

ACKNOWLEDGEMENT

THE quotations from the Bible in this publication are from the Revised Standard Version of the Bible, copyrighted 1946 and 1952 by the Division of Christian Education, National Council of Churches of Christ in the United States of America, and used by permission.

CONTENTS

LIST OF ILLUSTRATIONS

I

INTRODUCTORY

History and Religion

THE history and religion of Israel are inseparable and yet stand in sharp contrast to each other. The history of Israel is in one sense only a minor feature in the broad complex of ancient Near Eastern history. With the possible exception of the reigns of David and Solomon, Israel never attained imperial status. She was able to maintain fully independent national existence for only a fraction of the Old Testament period, and effective national unity for an even shorter time. She was involved in the great movements of history, not primarily because of her own national or imperial pretensions, but rather because she occupied territory across which one or other of the great powers sought to pass, or because she allowed herself to be used as a cats-paw to further the policies of one power against another.

The religion of Israel, on the other hand, has exercised an influence out of all proportion to Israel's political importance. From it have come two great world religions, Judaism and Christianity; and indirectly it has contributed much to a third, Islam. Its extraordinary vigour is already evident within the span of the Old Testament period. Whereas Israel's political influence was seldom extensive, and her national independence was often precarious or non-existent, her religion displayed a remarkable toughness in surviving crises, resisting or absorbing alien influences, and maintaining its own distinctive character far more successfully than the religions of neighbouring peoples, which were transformed, absorbed, or obliterated in the far-reaching changes by which the ancient Near East was affected.

In spite of this contrast, Israel's history and her religion are inseparable. It is, of course, true that wherever religion has existed it has been in some measure a factor in history and has itself been influenced by political, social, economic, and other factors. It is also a fact that in the ancient world of which Israel formed a part religion was not simply a department of life but permeated all human activities in a way which is less obviously true today. Religion was so intimately connected with other fields of experience that when, for instance, Israel settled in Canaan and adopted agricultural and urban ways of life, the change necessarily had a momentous impact upon religion. Again, whereas today the association of a small nation with a powerful one will probably have political and commercial consequences, in the ancient world there were often religious consequences as well: political subordination could involve religious influence, as in the reign of Manasseh, king of Judah; and political liberation could facilitate religious reform, as in the reign of Manasseh's grandson, Josiah. But in addition to these factors there is the historical character of Israel's own religion. Its foundation was the deliverance from slavery in Egypt, an event which left its stamp on Israelite worship, faith, and morals, and which also gave to Israel its strongest principle of national unity. Nor was it only in these early happenings that the God of Israel was thought of as acting. All through the Old Testament period outstanding events were interpreted in terms of God's purpose for His people. Indeed, the historical records which the Old Testament contains were compiled and given their present form not primarily to provide information about political, military, social, and economic developments, but to interpret the vicissitudes of Israel's life in terms of the divine purpose, to describe important religious developments and crises, and to drive home religious lessons. Thus these records both provide information about religion and are also themselves expressions of Israel's religious understanding of her own life. When with their help we try to reconstruct Israel's history, we cannot get away from Israel's religion; and, on the other hand, when we try to describe Israel's religion, we have to consider not only theological ideas and cultic institutions but also the march of events, the clash of social and political forces, and the changes through which the outward pattern of the national life passed. Accordingly, throughout this book the history and religion of Israel will be considered in close association with each other. Our subject is, in fact, the total life of Israel as it can now be reconstructed and understood from the ancient records.

The Sources

The books of the Old Testament are by far the most important sources for our knowledge of ancient Israel. The fact is worth emphasizing. It is often claimed that archaeological discoveries have shed light on the Old Testament. This is no doubt true; but it is also true that the Old Testament sheds light on archaeological discoveries, for many of them would be much harder to interpret if the Old Testament were not available to supplement them.

But the Old Testament records need to be handled with discrimination. They come from widely different periods; and most of them are complex works composed of diverse strands. It is, therefore, necessary to attempt to analyse them and to date their several parts. We may even venture on the more delicate enterprise of tracing the influences by which the various strands were fashioned before they were combined, or during the course of oral transmission before they were written down.

Further, it is important to recognize that the different elements in the material do not all have the same literary character. For example, the stories about Abraham, Isaac, and Jacob do not belong to the same literary class as either the stories about Adam and Eve on the one hand, or, on the other, the chapters in 2 Samuel about the history of David's court. The stories about Adam and Eve are not based on historical reminiscence or record, but are intended to depict and interpret man's relationship to his environment and to God; the patriarchal narratives are based on historical reminiscence and have a historical core which has been overlaid and in part transformed in the long process of transmission; the account of David's court may safely be regarded as a historical record, contemporary or nearly so with the events which it describes. The test of literary character is a necessary preliminary to the evaluation of historical reliability.

Again, the Old Testament records are remarkably incomplete. This is a characteristic which familiarity often leads us to overlook; but it is undeniable. Many questions which the historical inquirer would like to pose are left unanswered. There are some baffling gaps in the narrative. Although much space is devoted to some quite brief periods, others of greater length are described summarily. Probably something less than a century and a half is covered by 1 and 2 Samuel and 1 Kings 1–12; but 2 Kings 12–25 surveys the history of over two and a half centuries. The chief reason for this is that the Old Testament is not primarily a historical source book but a religious collection of writings. The writers paid attention to those periods, persons, institutions, and events which seemed to them to be of special religious importance. But the great and growing mass of archaeological material can help us to fill some of the gaps and to understand better the evidence which the Old Testament supplies.

Where the incompleteness is the result of deliberate selection it is one means by which the writers express their interpretation of Israel's historical experience. As any historian is bound to do, they have discarded material which seemed to them to be of less importance. There are other ways in which this interpretative element appears; and in using the documents for the purposes of historical reconstruction it is important to recognize and assess the writer's point of view, which may be implicit in the way he tells the story or explicit in the comments which he makes on it.

There are in the Old Testament three main blocks of narrative material which are relevant to the task of historical reconstruction. The first is the sequence of stories (interspersed with codes of law) relating to early times, which we find in the first few books of the Bible. Although there are strong traditional and other reasons for treating the first five books (*The Pentateuch*) as a coherent group, the book of Deuteronomy is probably best taken not as the coda to what precedes but as the preface to what follows. This leaves the first four books, *The Tetrateuch*, as they are now sometimes called, as the initial group. In them three main sources are interwoven. The final work of editing them was done after the Exile; but they embody much ancient traditional material. The second block consists of Joshua, Judges, 1 and 2 Samuel, and 1 and 2 Kings, with Deuteronomy as the preamble to the whole. Here a rich variety of material has been woven together: traditional tales, official or semi-official annals and lists, eyewitness accounts of important national events, stories derived from prophetic groups, and the like. While there is no rigid uniformity of approach, the general principle of interpretation and the standards by which persons and policies are judged are those of the book of Deuteronomy. This compilation, spanning the period from the entry into the Promised Land till the Exile, is now commonly known as *The Deuteronomistic history*. The third block is *The Chronicler's history*. It consists of 1 and 2 Chronicles, Ezra, and Nehemiah. Formally it covers the entire period described by the other two compilations and goes beyond them. But from Adam to Saul there are only genealogies and lists, with no accompanying narrative. The account of the monarchic period is based on the books of Samuel and Kings, with additions and extensive omissions. Of the life of the post-exilic community in Judah there is only a fragmentary record, which, though it includes documents of great interest and value, also raises perplexing problems, particularly of chronology. This presentation of Israel's history is characterized by the outlook of the post-exilic Jewish community, its cultic interests, and its relations with the Samaritans.

Useful evidence about the life of ancient Israel can also be gleaned from other types of literature in the Old Testament. The books of the prophets, some of which themselves contain narrative material, provide much information about social and religious conditions, which is in some sort a commentary on the historical books. The compilations of the laws, and the Psalms, also provide their own quota of evidence, though it should be noted that in the Psalter there are particularly difficult problems of interpretation and dating.

The biblical evidence can be supplemented by the results of archaeological investigation. The material available is, in the main, of three kinds. First, there are the remains of buildings, sacred and secular, and a wide variety of objects used in worship and everyday life. Second, there are representations of gods, men, and animals, in wall-paintings, reliefs, statues, and the like. Third, there are documents: inscriptions on stone or metal, clay tablets, and papyri. Such material can be of great help to the student of the Old Testament; but we must be careful not to misunderstand or exaggerate its importance. The evidence has to be *interpreted*. This is true not only of texts but also of the other kinds of material just mentioned. It may not be immediately obvious what purpose was served by a utensil, an article of temple equipment, or even a building; and the identity of the persons represented in paintings or sculpture may not be unmistakably evident. The evidential value of archaeological data must also be critically assessed. The fact that the inscription of an Egyptian Pharaoh or an Assyrian king is contemporary with the events which it records, whereas the Old Testament account of the same events was composed or edited at a later date and is preserved in manuscripts which are many centuries later still does not mean that the former may be uncritically used to correct the latter. Contemporary documents may contain accidental errors or even deliberate falsifications. Again, archaeological data are collected, observed, recorded, and interpreted by archaeologists, who, like biblical critics, are neither infallible nor unanimous in all their views. Techniques are improved; theories are modified; systems of chronology have to be revised. Accordingly, the science of archaeology should be regarded neither as providing comprehensive corroboration of the biblical record nor as an infallible corrective to it. Archaeological evidence helps to solve some problems; but it also appears to complicate others. Probably its greatest services to Old Testament study are to provide us with a clearer framework of international history in which to set the history of Israel, and to help us to recapture the atmosphere of the world in which ancient Israel lived its life.

The Geographical Setting

The land which played so important a part in the religious aspirations of Israel exercised a considerable influence on the nation's life. Some knowledge of its main features and of its relationship to neighbouring territories is essential for any understanding of the Old Testament records.

The position of the land of Israel between the great centres of civilization in the ancient Near East made it a much frequented thoroughfare for traders and armies and an area open to varied cultural and religious influences. Three regions call for special consideration.

1. To the south-west lay Egypt. Its population was concentrated in the Delta and along the narrow strip of fertile territory on either side of the Nile. In the south it was protected by the Nile cataracts and on the west by the desert. The hazardous north-eastern approaches across the desert of Sinai provided the principal connexions with western Asia; and when Egypt was strong it was by this route that its armies advanced, traversing Palestine before they could reach any more distant objective.

2. Eastward, beyond the desert, lay the other great riverine centre of civilization, the region between the Tigris and the Euphrates, which was much more exposed than Egypt and was at various times invaded from the mountains in the north and east and also from the west. From early times various peoples (Sumerians, Akkadians, Amorites, Kassites, Assyrians, Babylonians, Persians) dominated part or all of this region; and from it some of them pushed westward towards Asia Minor, the Mediterranean seaboard, or Egypt. Some of these expansionist thrusts inevitably affected Israel, which lay at the western extremity of what in modern times has been called 'the Fertile Crescent', the great sweep of fertile territory which extends from the head of the Persian gulf, along the Tigris–

Euphrates region, westward to Syria, and south towards Egypt.

3. North of Palestine are the territories which we know to-day as Syria and Lebanon, and beyond them is Asia Minor, anciently the seat of Hittite power, from which any southward movement would be a threat to Palestine.

Much of the political and military history of Palestine was determined by conditions in these three regions. If all three were weak, the opportunity was given for a local power to dominate Palestine and its immediate environment. If the rulers of one or more of them pursued expansionist or aggressive policies, Palestine became at best a buffer state and at worst a cockpit.

The name 'Palestine' is derived from 'Philistia', i.e., 'the land of the Philistines.' What was originally the name of the south-western region came to be applied to the entire land which in the Old Testament is generally called 'the land of Canaan' or 'the land of Israel'. There are four main geographical divisions, each of which runs from north to south.

1. In the west is the coastal plain, lying behind a coastline which, unlike that of Phoenicia further north, has no good natural harbours. There is considerable variation in the width of the plain. The jutting ridge of Carmel narrows it to some 200 yards; but to the south it broadens into the plain of Dor, the plain of Sharon, and, beyond Joppa, the plain of Philistia. This whole strip of territory provided an easy line of communication from which other routes branched in a north-easterly direction through the hill passes south-east of Carmel.

2. The backbone of the country west of the river Jordan consists of three blocks of hill country. In the extreme north are the highlands of Galilee, bounded on the south by the great plain of Esdraelon. In the centre is the hill country of Ephraim, broken up by broad, fertile valleys. It is difficult to draw any clear line of demarcation between this region and the hill country of Judah in the south; but certain marked differences of character are obvious. The Judaean hills are less fertile and more closely set. Between them and the plain lies

the Shephelah, a low ridge which forms an outlying bulwark between the hill country and the maritime plain, whereas further north access to the central highlands is more direct and easy. In the far south lies the Negeb (or Negev), a relatively barren region merging into the desert.

3. The most striking feature in the entire country is the great cleft which lies between the western and eastern hills. Through it flows the river Jordan, from its sources near the foot of Mount Hermon, through Lake Huleh to the Sea of Galilee, where it is 695 feet below the level of the Mediterranean, and then on to the point at which it enters the Dead Sea. The distance between these two inland seas is some 65 miles as the crow flies; but the river pursues a tortuous course of about 200 miles and drops to 1,285 feet below the level of the Mediterranean. South of the Dead Sea lay the land of Edom.

4. Although the term 'Canaan' was applied to the territory west of Jordan, a survey of the Holy Land would be incomplete without some account of Transjordania. North of the river Yarmuk was Bashan, rich in corn and cattle, and south of it Gilead. Further south were the kingdoms of Ammon and Moab, the former south of the river Jabbok, with its capital at Rabbah, or Rabbath Ammon (the modern Amman), the latter south of the river Arnon.

It has been said with justice that Palestine has no obvious natural frontiers except the Mediterranean coastline. No sharply defined geographical features separated it from neighbouring territories. On the other hand, its general character made difficult the establishment and maintenance of internal political unity. The contrasts between hill country and plain, and the separation of different hilly regions by such features as the plain of Esdraelon and the Jordan valley fostered the limited local loyalties which are a recurring phenomenon in Old Testament history.

The geographical variety of the land makes for a considerable range of fertility and climatic conditions. In Old Testament times the land was mainly agricultural (corn, grapes, and olives being the chief produce), with an influential urban

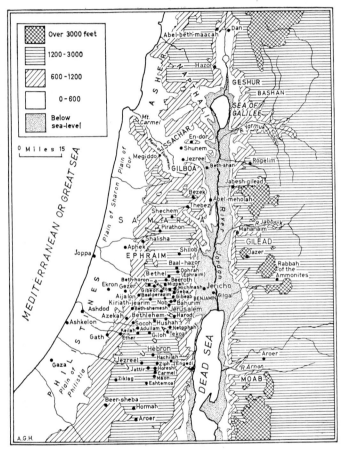

Map 1. The Holy Land.

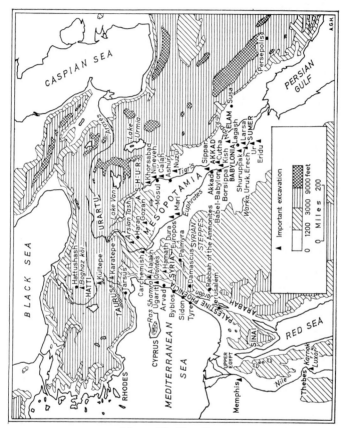

Map 2. The Near East.

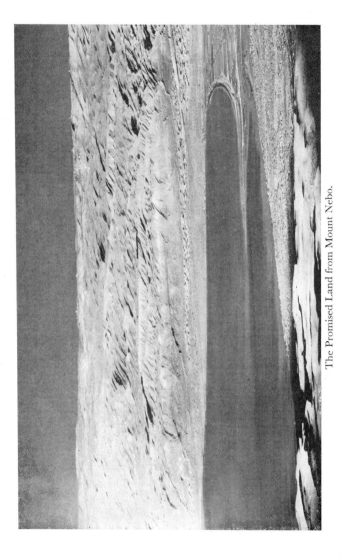

The Promised Land from Mount Nebo.

element, and with considerable pastoral fringes. The two main seasons are the hot, dry summer and the rainy season. The agricultural year began with the autumnal rains of October and November (the former rain) which prepared the ground for ploughing and sowing. At the end of the rainy season, in March and April, came the showers (the latter rain) which were needed to swell the grain and ensure an ample harvest. First the barley was garnered, then the wheat; and at the end of the summer the ripened fruits were brought in and the annual cycle began afresh.

Such was the land in which the Israelites settled. For most of the Old Testament period their occupation of it was only partial; for long periods they occupied it only as the vassals of foreign powers; and always their life was in varying degrees interpenetrated by other peoples. According to their own traditions their ancestors came into it as aliens and the foundation of their religion took place beyond its borders. Yet it became one of the most expressive symbols of their national religious hope.

II

THE PATRIARCHS

WHEN did Israel's history begin? It could be argued that the establishment of the monarchy is the appropriate starting-point, since it was then that the tribes were welded into an organized state, and then, too, that the serious compilation of historical records seems to have begun. Or the settlement of the tribes in the land of Canaan might be suggested, since at this point the confederate tribes are first clearly seen to have a common life in their own land. Or one might go further back still to the Exodus from Egypt as the decisive sequence of events which gave to the Israelite tribes their religious and national unity. The Old Testament takes us still further back. It represents the people of Israel as derived from a common ancestor, Jacob, whose name was changed to Israel (Gen. 32^{28}, 35^{10}); and as a preamble to the Exodus story it records groups of narratives about Jacob's grandfather and father, Abraham and Isaac, and about Jacob himself and his twelve sons (Gen. $12-50$).

These stories have traditionally been regarded as historically accurate records of events. At the other extreme they have been treated as giving little or no reliable information about the period to which they refer, but as reflecting the religious belief and practice, and the folk traditions, of the period of the monarchy. But on any showing they are of historical importance since at the very least they embody Israel's own understanding of her remoter past. To determine more precisely their value as sources and to appropriate the evidence which they provide for our present purpose it is necessary to ask three questions: (1) What is the character of these stories and how are they to be interpreted? (2) How does their evidence accord with what is known from other sources about the period to which they relate? (3) What information do they provide about pre-Mosaic Hebrew religion?

1. The traditional view that the patriarchal stories are in every detail accurate records of events and conversations is extremely difficult to maintain on grounds of general probability and in face of the literary evidence which shows that sources from different periods have been interwoven in them and that these sources sometimes give varying accounts of the same events.

At the other extreme is the theory that some at least of the narratives were originally myths about gods, and that the divine figures have been scaled down to human proportions. In support of this two main arguments have been advanced. On the one hand, similarities have been pointed out between some of the names in the patriarchal stories (e.g. Gad, Asher, Terah, Milcah, Laban) and divine names found in other ancient Near Eastern sources. On the other hand, it has been supposed that the association of individual patriarchs with particular sanctuaries (Abraham with Hebron, Isaac with Beersheba, Jacob with Bethel) points back to a stage at which they were the gods of these holy places. But the evidence from these names is highly precarious; and the entire theory is so much at variance with the spirit of the narratives that it must be rejected.

There is much more to be said for the view that at least sometimes the individual personages in the stories represent tribes or other communities. This is in accord with the concept of 'corporate personality', which plays an important part in the Old Testament. The life of a community is conceived of and described in vividly individual terms; and, on the other hand, a representative individual can embody the life of the community in what he is, does, and experiences. There are clear instances of this mode of thought in the oracle to Rebekah (Gen. 25^{23}), the blessings on Jacob and Esau (Gen. $27^{27-29, 39-40}$), and Jacob's poetic utterance about his sons (Gen. 49^{1-27}), where what is said to or about individuals relates in fact to the character and history of tribes or peoples. It is also reasonable to suppose that Gen. 31$^{44-49}$ describes the settlement of a border dispute between the Israelite and Aramaean peoples. There are other stories in which a collective rather than an individual interpretation fits very well; but it is by no means appropriate to all the patriarchal narratives.

As we have already seen, some of the stories connect one or another of the patriarchs with this or that sanctuary. These, no doubt, were recounted as part of the sacred tradition of the sanctuaries before being worked into the record as we now have it, or even into the sources of which it is composed. Some of them may have been told to account for the origin of the sanctuaries in question. Akin to these are the aetiological stories, which explain the origin of a custom (e.g. Gen. 32^{32}), the bestowing of a name (e.g. Gen. 21^{31}), or the existence of some natural feature (e.g. Gen. 19^{26}).

Thus the variety of the material is such that no single interpretative principle can satisfactorily be applied to all the stories. They include both collective and individual elements. The most appropriate comprehensive term to apply to them is *saga*, since in them we find neither pure fiction nor scientific history, but folk traditions which deal, or purport to deal, with actual events. To assess their historical value it is necessary to examine the archaeological evidence.

2. Here the first problem which confronts us is a chronological one, since if we are to consider extra-biblical evidence bearing on the patriarchal age we must have at least a rough idea of where the patriarchs fit into the framework of ancient Near Eastern history. The patriarchal stories by themselves afford little help. The patriarchs occupy the centre of the stage; contemporary powers and rulers are very much in the wings; and references to them are not historically or chronologically precise. Two points have been taken to indicate the beginning and end of the period. The first is the reference to Abraham's victory over the four kings (Gen. 14). The name 'Amraphel' (v.1) used to be equated with 'Hammurabi' and regarded as a reference to the celebrated king of Babylon, who was thought to have lived about 2000 B.C. The second is the descent of

Jacob and his family into Egypt (Gen. 46). During the period 1720–1550 B.C. Egypt was dominated by Asiatic invaders known as the Hyksos; and it has generally been supposed that this provides the most appropriate period for the Hebrew migration. The patriarchal period would then be roughly the first half of the second millennium B.C. But the equation of Amraphel with Hammurabi is now generally rejected; and the identification of the other kings mentioned in Gen. 14 is beset by difficulties. Moreover, the archaeological material now available has made necessary an extensive revision of the chronology of western Asia at this period; and Hammurabi's reign is now commonly brought down to about 1700 B.C. It is also by no means certain that the descent into Egypt took place during the Hyksos period. As we shall see when we consider the date of the Exodus, a strong case can be made for the view that the sojourn in Egypt began in the first half of the fourteenth century B.C. This leaves us with a somewhat extensive period in which to look for relevant archaeological evidence. Material of special interest has come from two places: Mari (Tell Hariri) on the middle Euphrates, and Nuzu (Yorghan Tepe) near the modern Kirkuk.

During the early part of the second millennium B.C. various areas in Mesopotamia and even in Syria and Palestine came under the domination of Semitic invaders who are generally known as Amorites (though the appropriateness of the term has been questioned). One of the most important of their centres was Mari. Since 1933, excavations there have brought to light thousands of clay tablets which provide valuable information about the history of Mesopotamia in the nineteenth and eighteenth centuries B.C., leading up to the time when the city was taken and destroyed by Hammurabi of Babylon. It appears that Mari's sphere of influence included Haran, from which, according to Gen. 12⁴, Abraham migrated to Canaan. The names of some of Abraham's relatives correspond to the names of cities in the vicinity; and in general the patriarchal names resemble 'Amorite' names found in the Mari texts. There are references to a marauding tribe of Benjamites

(*Banu-yamina*).[1] From such evidence it has been argued that the Mari texts provide the appropriate background for some of the tribal movements to which the patriarchal narratives bear witness.

The Nuzu texts give a vivid picture of social and economic conditions in a town in the kingdom of Mitanni during the fifteenth century B.C. The ruling class was Indo-Aryan; but the bulk of the population consisted of Hurrians, a non-Semitic people who since the beginning of the second millennium had infiltrated into many parts of the Fertile Crescent, including Syria and Palestine. In the Old Testament the Hurrians appear as 'Horites', for which 'Hivites' is an erroneous equivalent. Several of the customs and legal usages provide striking parallels to features in the patriarchal stories: e.g. the adoption of an heir by a childless couple; the reversion, in such circumstances, of the right of inheritance to a son born subsequently; the provision by a barren wife of a concubine for her husband and the law against expelling such a concubine and her children; the adoption of a son-in-law where there were no sons in a family; the fact that possession of the household gods constituted the title to the inheritance.[2] The essential point is not the mere existence or number of the parallels, but the fact that these usages are not found in the records of Israelite life after the settlement in Canaan. It is, therefore, evident that the patriarchal narratives do not merely reflect conditions, practices, and beliefs in Israel in the period of the Judges and the monarchy, but have preserved faithfully the traditions of a much earlier age.

In fairness it must be recognized that there is a chronological gap of some three centuries between the Mari and Nuzu texts, though the usages to which the latter bear witness must have been older than the texts themselves. It must also be acknowledged that the evidence relating to the patriarchal stories which is provided from these and other sources is of a quite general kind. As yet no discovery has enabled us to date with anything approaching precision any event recorded in Gen.

[1] See *ANE*, p. 260.   [2] See *ANE*, pp. 167–9.

12–50; nor have we any archaeological evidence to show con-clusively that any of these events did in fact take place, or that any of the persons mentioned did in fact live. But it is now clear that the stories about Abraham, Isaac, and Jacob reflect with remarkable faithfulness the conditions of a period much earlier than that in which they were given literary form. (The cycle of stories about Joseph will be discussed in connexion

Asiatics visiting Egypt.

with the Exodus.) Though they contain some anachronisms, there can be no doubt that they embody reliable traditions of great antiquity. The corroboration which they have received should make us cautious about rejecting any parts of them as unreliable unless there are good grounds for doing so. There are no good grounds for denying that the patriarchs were individual persons who actually lived, though some of the stories told about them may refer to events involving communi-ties rather than individuals. On the other hand, as we have seen, the corroboration is quite general; and there is no specific evidence of their existence apart from the biblical record.

3. The picture of the patriarchs which Genesis gives us is of semi-nomads moving from one part of the Fertile Crescent to

another, and sometimes even venturing as far as Egypt in search of food and pasture. 'A wandering Aramaean was my father' (Deut. 26⁵. R.V., 'A Syrian ready to perish was my father'): so began the formula which the ancient Israelite used when he brought his first-fruits to the sanctuary, thus preserving the tradition that his ancestors had belonged to an Aramaic-speaking group of this kind.

Texts from widely separated periods and from different parts of the ancient Near East show that in such communities it was common for the cult to be traced back to a special relationship between the deity and the clan chief or cult founder. In a cult of this kind the relationship of the deity was with the community and its chief or ancestor. It was personal in character and lacked the prominence given in Canaanite religion to the association between the god and the land or a specific locality. Further, it was by an act of choice that this close relationship was established between the cult founder and the god. It is, therefore, entirely appropriate that in the patriarchal stories we find such a relationship expressed in the titles, 'the God (or Shield) of Abraham' (Gen. 15¹, 28¹³, 31⁴², ⁵³), 'the Fear (or Kinsman) of Isaac' (Gen. 31⁴², ⁵³), and 'the Mighty One of Jacob' (Gen. 49²⁴). It has been maintained that these three titles indicate the existence of three separate cults which were later fused in the worship of the Mosaic Yahweh, with whom the gods of the fathers came to be identified. Even if this is rejected, it is clear that the religion of the patriarchs as described in Genesis has a personal character in both its individual and communal aspects which is in accord with the situation of the patriarchs, and which marks it off as different from agricultural fertility cults and also from the state cults of the great powers. This personal character remains an obstinate feature of Old Testament religion throughout its entire development.

Other divine names or titles in the patriarchal stories deserve attention. In Exod. 6³ it is stated that God revealed Himself to the patriarchs as El Shaddai (traditionally translated as 'God Almighty'). This name occurs in Gen. 17¹, 28³, 35¹¹. The original meaning of 'Shaddai' was perhaps 'He of the

Mountain'; but this is not wholly certain. The name 'El Elyon' (God Most High) appears in the story of Abraham's encounter with Melchizedek (Gen. 14[18 ff., 22]), where there is an explicit reference to God as Creator and a probable reference to Jerusalem, with which Salem should almost certainly be equated. It has been plausibly suggested that Elyon was an ancient divine title in use at the sanctuary there before the Israelite settlement in the land. There are also references to El Roi (God of Seeing?; Gen. 16[13]), El Olam (the Everlasting God; Gen. 21[33]), and El Bethel (the God Bethel; Gen. 31[13], 35[7]). The word 'El', which appears in all of these, is the common Semitic term for 'god'. It is also the name of the supreme deity in the Canaanite pantheon. It has been inferred that the above titles referred in pre-Israelite Canaanite usage either to separate local deities or to different local manifestations of the one supreme El. The latter seems the more probable supposition. In the texts discovered at Ras Shamra in northern Syria there are several instances of such compound titles used in this way.

As we shall see, the settlement of the Israelite tribes in Canaan was probably both a complicated and a protracted process. It is, therefore, very difficult to determine how and when such elements from the religion of the land came to be embodied in the patriarchal traditions. The same may be said of the references to sacred sites with their characteristic features, such as sacred trees (Gen. 12[6], 13[18], 14[13], 21[33], 35[4]), the sacred spring (Gen. 16[14]), and the sacred pillar (Gen. 28[18], 35[14]). Combined with the ancient records of the life and faith of the semi-nomadic forefathers there are reflections of Israel's encounter with the religion of the land. But all the varied elements have been included within a comprehensive framework, which makes the sequence of stories one story and gives to that story its present meaning and character, as the prelude to the affliction of Israel in Egypt, the deliverance from bondage there, and the making of the covenant between God and His people, and as the promise that God will give to the descendants of the patriarchs the land of Canaan as their inheritance.

III

EXODUS AND CONQUEST

The Story

THE theme of a deliverance from bondage in Egypt and a subsequent entry into the land of Canaan not only provides the framework of the opening books of the Old Testament but is also recalled and emphasized in many other passages. Where it is not directly referred to, it is often the implicit presupposition of what is said. There can be no reasonable doubt that this theme was of prime importance both for the national self-consciousness and for the religious self-consciousness of Israel. They were bound to each other as a confederation of tribes and as a people not simply by the ties of a common descent but by the experience and the consequences of a common deliverance, and by a covenant by which their God had united them to Himself and to each other. Their relationship to their land was not that of aboriginal settlement, nor even based on any supposed right of conquest: the land was theirs by divine promise and gift; and when many of them were driven into exile they thought of restoration as a new fulfilment of the divine promise accomplished by a new act of divine grace comparable to the first.

Nor can there be any reasonable doubt that this recurring theme corresponds in some sense to an actual sequence of historical events, and that the Exodus and the conquest did take place. It has often been observed that if we had had no record of Moses we should have had to presuppose the existence of such a person to account for what was accomplished through him; and it has also been argued that the Exodus story is in a measure authenticated by the unlikelihood that any people would have invented the account of the humiliating bondage of their ancestors in Egypt. Whatever degree of cogency we may allow to such contentions, the fact remains that the theme of the Exodus and its consequences is too deeply ingrained in the Old Testament records to be written off as mere fiction, though there is no ancient evidence independent of the Bible which directly corroborates it.

Granted, however, that actual historical events lie behind the Old Testament traditions about the Exodus and the conquest, it is difficult to date those events and to determine their precise character and scope. Much modern research has been devoted to the problem of chronology; but this cannot be separated from a consideration of the early relationship and organization of the Hebrew tribes. We are obliged, in fact, to consider not one but several interlocking historical problems; and the evidence available for their solution, whether from the Bible or from archaeology, is far from straightforward.

At first sight the narrative in Scripture is tolerably plain: the descent into Egypt during Joseph's lifetime; the subsequent reversal of Egyptian policy towards the Hebrews; the call of Moses while in exile to return to Egypt to lead his people out; the infliction of the plagues on Egypt and the alternation of willingness and reluctance to let the people go which the Pharaoh displayed; the final blow of the tenth plague; the hasty departure and the deliverance at the Red Sea; the journey to the Mount of God, where the Covenant was made and the Law given; the vicissitudes of the wilderness wanderings, prolonged to forty years because of the people's complaints and unbelief; the final crossing of the Jordan and the rapid conquest of the entire country.

Places and Times

Even geographically there is much uncertainty about the details of the narrative. The 'Red Sea' is in Hebrew *yam suph*, 'the Sea of Reeds'; and general considerations make it likely that the crossing took place not at the head of the Gulf of Suez, which seems too far away from the Hebrew's point of departure, but at one of the lakes now joined by the Suez Canal. Other suggestions are the head of the Gulf of Akaba (referred to as *yam suph* in 1 Kings 9²⁶), though this seems altogether too remote, and Lake Sirbonis, which is at variance with Exod. 14² ᶠ·. The Mount of God, which the Hebrew sources variously designate Sinai and Horeb, has been traditionally identified with Jebel Musa in the south of the Sinai. Against this it has been argued that this would have taken the Israelites dangerously near to the route which the Egyptians used to reach the copper and turquoise mines in that region. Another view is that the mountain lay in the northern part of the peninsula, not far from the oasis of Kadesh; but although there are some indications that the tribes may have journeyed directly to Kadesh on leaving Egypt (cf. Judges 11¹⁶), it is stated elsewhere (e.g. Deut. 1²) that Sinai and Kadesh were far apart. Again, the account of the happenings at the mountain has seemed to suggest volcanic activity (though this is an unnecessary inference); and since no mountain in the Sinaitic peninsula has been volcanic in historical times, and for other reasons, it has been argued that the site lay in north-western Arabia, east of the Gulf of Akaba. Though the traditional location has perhaps more to be said for it than has sometimes been supposed, the Old Testament records do not provide clear evidence on the subject; still less do they enable us to trace the route followed across the wilderness.

The evidence bearing on the date of the Exodus is more plentiful but difficult to evaluate. Four related questions call for an answer: (1) When did the Hebrews enter Egypt? (2) How long did they stay there? (3) When did they leave? (4) When did they occupy the land of Canaan?

1. In considering the chronology of the patriarchal stories we have already noted (see above, p. 17) the view that the period of Hyksos domination would have been a favourable time for Joseph's promotion and for the settlement of the Hebrew clans in Egypt, since Asiatics would presumably have been more welcome then than during the periods of native Egyptian rule. There is, however, Egyptian evidence that at other periods tribesmen from the desert were allowed to enter the eastern region of the Delta, and that Semites sometimes

held high office in the Pharaohs' service. It might be thought that the names and other Egyptian details in the Joseph story would help to determine the date more precisely; but they do not. The personal names (Potiphar, Asenath, Zaphenath-paneah) are characteristic of the tenth century. The reference to a chariot (Gen. 41⁴³) merely precludes a date earlier than the Hyksos period. It is improbable that a Hyksos king would have given the daughter of the priest of On (Heliopolis) to Joseph as his wife (Gen. 41⁴⁵), since the Hyksos despised the sun-God Ra, whose great temple was at On. Such items of local colour as there are do not point to any definite period.

2. The duration of the sojourn is given in Gen. 15¹³ as 400 years and in Exod. 12⁴⁰ f. as 430 years. In the latter passage the Greek and Samaritan texts make this period cover in addition the preceding patriarchal sojourn in Canaan. This may be an attempt to reconcile this long period with indications that four generations spanned the entire sojourn (Gen. 15¹⁶; Exod. 6¹⁴⁻²⁷). However that may be, it can be argued that the genealogical evidence points to a sojourn considerably shorter than four centuries, perhaps barely a century and a half.

3. The date of the Exodus appears to be precisely indicated in 1 Kings 6¹ as 480 years before the founding of the Temple in the fourth year of Solomon's reign. If Solomon came to the throne in 961 B.C., this would date the Exodus in 1438 B.C., during the eighteenth Egyptian dynasty. But it has been argued that 480 is an artificial figure (40 × 12), and that the date conflicts with the statement in Exod. 1¹¹ that the Israelites 'built for Pharaoh store-cities, Pithom and Raamses'. The sites of these cities have been identified. Raamses, which had been Avaris, the Hyksos capital, was rebuilt and restored as the capital in the thirteenth century by Rameses II, a Pharaoh of the nineteenth dynasty, who also carried out building work at Pithom. Scholars who have preferred the earlier date for the Exodus have alleged that Exod. 1¹¹ is an interpolation.

4. In relation to the scale of Egyptian history, the arrival, sojourn, and departure of the Hebrew tribes would be minor

events. It is, therefore, not surprising that contemporary Egyptian records provide no specific reference to any of them. Even if one recognizes that happenings like the plagues and the destruction of the Egyptian army in the Sea of Reeds were of national importance, their humiliating character might have kept them out of official records. On the other hand, an invasion of Palestine such as is described in the Book of Joshua might well have left some trace in written records and in the remains of cities destroyed. The tablets found at Tell el Amarna in Egypt in 1887 contain letters sent from the Egyptian dependencies in Palestine and Syria appealing for reinforcements. There are references to invaders called Habiru.[1] This name is found in other documents from various parts of the Near East during the second millennium. It seems to denote not a specific ethnic unit but loosely organized groups of landless people. It has been equated with 'Aperu', a name frequently applied in Egyptian monuments to groups serving as labourers or mercenaries. If, as some have maintained, these names may be philologically related to 'Hebrews', then the biblical Hebrews must have been only a part of the Habiru. Not unnaturally, however, the references in the Amarna letters to the Habiru have been taken as describing from the defenders' side the invasion under Joshua. This would date the invasion c. 1400 B.C., implying for the Exodus a date in the middle of the fifteenth century, and the Hyksos period as the probable beginning of the sojourn in Egypt.

In corroboration of this, Professor J. Garstang claimed to have shown from the results of his excavations at Jericho that the city fell c. 1400 B.C. But other cities, such as Lachish, said to have been taken by the Israelites appear to have been destroyed towards the end of the thirteenth century. Again, the biblical narratives say (Num. 20[14-21]; Deut. 2[1-9]) that when the Israelites were approaching Canaan they had to avoid the territory of the Edomites and Moabites; but archaeological surveys have revealed no evidence of sedentary occupation of that region for centuries before 1300 B.C. This strengthens the

[1] See *ANE*, pp. 262–77; *DOTT*, pp. 38–45.

Brickmaking in Egypt.

case for a thirteenth-century date for the invasion. To such evidence it must be added that Dr. Kathleen Kenyon's work on the site of Jericho (1952–8) has shown that the city which Garstang supposed to have been destroyed by Joshua fell *c.* 1580 B.C., which is much too early. Of Joshua's Jericho nothing appears to have survived to enable us to date its fall.

One further piece of archaeological evidence must be noted. On a victory stele of Pharaoh Merneptah, the successor of Rameses II, there occurs the earliest extra-biblical reference to Israel. This is in a list of conquered regions and peoples in Palestine;[1] and since Israel is named as a people rather than as a country, it has been inferred that the Israelites were not yet properly settled in the land, and therefore had comparatively recently entered it.

On balance it seems most likely that the conquest under Joshua took place towards the end of the thirteenth century and the Exodus earlier in the same century. If the sojourn lasted 400 or 430 years (Gen. 15^{13}; Exod. 12$^{40\,f.}$), the descent into Egypt must have taken place during the Hyksos period; otherwise, following the genealogical evidence, it may be supposed to have taken place during the first half of the fourteenth century, possibly in the reign of the heretic Pharaoh Ikhnaton. But neither the biblical nor the archaeological evidence presents a tidy picture; and there are some features which suggest that the chronological problems must be considered in conjunction with the question of the nature and scope of Israel's settlement in Canaan.

Israel's Occupation of Canaan

The difficulty of fitting together all the items of biblical and archaeological evidence relating to the Egyptian sojourn, the Exodus, and the conquest of Canaan has been taken to indicate that the sequence of events was more complicated than a superficial reading of the biblical narratives suggests.

[1] See *ANE*, p. 231; *DOTT*, pp. 137–41.

1. There may have been an extensive settlement before the Egyptian sojourn. The patriarchal stories tell how individuals with their immediate kin settled temporarily in various parts of Palestine. But we have already noted (see above, p. 15) the possibility that sometimes at least the individual may represent a community. The account in Gen. 34^{25} $^{ff.}$ of the attack on the Shechemites by Simeon and Levi is more readily understood if these two names represent not individuals but clans. This seems to be borne out by Gen. 49$^{5-7.}$. Such features point to a settlement in considerable numbers involving warlike methods (cf. Gen. 48^{22}) before the Egyptian sojourn; and the plausible suggestion has been made that this may be part of the movement to which the Amarna letters refer. It then becomes possible to think of Hebrews as forming part of the Habiru who penetrated into Palestine early in the fourteenth century, and still hold to a thirteenth century date for the invasion under Joshua.

2. It is widely held that only some of the tribes which formed the later Israel went down into Egypt. The evidence is somewhat elusive. It is true that the Bible describes a relatively small migration: Gen. 46^{27}; Exod. 1^5; Deut. 10^{22} speak of only seventy persons at the beginning of the Egyptian sojourn. The ancestors of all twelve tribes are mentioned; but it is claimed that this merely reflects a desire to include in the decisive experience of the sojourn and the Exodus all the main components in the later Israel, and further that Joshua 24 describes a covenant ceremony in which Hebrews who had not gone down into Egypt were united (perhaps together with Canaanite elements) in allegiance to Yahweh the covenant God of those tribes whom Joshua had led across Jordan.

3. The biblical accounts of the conquest cover four main areas: Transjordania, the central hill country, the southern region, and the north.

The occupation of Transjordania was made possible by the victories over Sihon and Og, but according to the narratives (Num. 32; Deut. 3$^{18-20}$; Joshua 1$^{12-15}$, 22$^{1-6}$) it was not effectively completed until after the country west of Jordan had been taken.

The advance of the Israelites into the other three regions is described in Joshua 1–11. At first sight these chapters seem to present a picture of a conquest which was swift, complete, and carried out by the united action of the whole of Israel. But there are indications later in the book that substantial areas remained under Canaanite control and that in places the Israelites were subject to the Canaanites (Joshua 13$^{2-6}$, 15^{63}, 16^{10}, 17$^{11-13}$). This is what might be expected, since we hear in later periods of the corrupting influence of the Canaanites on Israel. Moreover, Judges 1 tells of regional operations by individual tribes or groups of tribes. This appears to be an ancient narrative, giving a faithful account of some phases of the occupation. There are also references in Num. 14$^{44 f.}$ and 21$^{1-3}$ to attacks made from the south; and it is commonly inferred that these point to a movement separate from Joshua's campaign.

But although we must allow for a complex process of occupation and for the probability that the narratives in the first half of the book of Joshua have been somewhat simplified and streamlined, it is going too far to break the record up, as has sometimes been done, into separate *aetiological* stories (i.e. explanations of local landmarks, customs, and the like) and deny to it any real cohesion or unity. The sequence of events is coherent and credible. Striking first at Jericho, the Israelites follow up their initial success with an ill-fated assault on a city to the north-west which the narrative calls Ai, but which may have been the nearby Bethel, since Ai was not occupied at this time. A second attack was successful. It was followed by the securing of the Gibeonite cities against the attack of five kings from the south, who might otherwise have prevented the Israelites from penetrating further into the country.

Thus far, important bases had been secured for the more extensive occupation of the central region of the land. About the accounts of two subsequent phases in Joshua's campaign some questions arise. The description of his operations in the south (Joshua 10$^{28-39}$) is difficult to reconcile with what is recorded in Judges 1 about the capture of cities in that region.

In the far north Joshua is said to have destroyed the important city of Hazor (Joshua 11$^{1-15}$). It has been argued that it would have been difficult for Joshua's forces to strike north from the central hill country across the plain of Esdraelon, the southern limit of which was guarded by a line of fortified cities. But the difficult is not necessarily the impossible; and excavation of the site of Hazor has brought to light evidence that the city was destroyed in the latter part of the thirteenth century. This seems also to have been the period at which some of the southern cities were destroyed, such as Lachish, Debir, and Eglon. Thus there is a considerable amount of archaeological evidence which is in harmony with the biblical records. But we must be careful not to press this evidence too far. It does not by any means provide us with a coherent picture of the conquest of the land. The basic evidence for the conquest and settlement is still the biblical record. The narrative in Joshua forms part of the Deuteronomistic history (Joshua–2 Kings), which lays great emphasis on the unity of all Israel and is hostile to Canaanite influence. Here, as in other books, it embodies earlier traditions and documents; but it has so presented them that the campaign appears as a swift and unified process, resulting in the removal of all Canaanite power.

The Foundations of Israel's Religion

The many references elsewhere in the Old Testament to these events make it clear that later generations looked back to them not only as the creation of the confederacy of Israelite tribes and their establishment on Canaanite soil, but also, and primarily, as the foundation of Israelite religion. To them Israel's faith, worship, and obedience were all related. It is, therefore, important to inquire, what may be known or reasonably inferred about Mosaic religion.

The story in Exod. 3 of the call of Moses implies that it was on that occasion that the name of Israel's God was revealed (vv. 13 f.). This is made more explicit in Exod. 6$^{2 \, f \cdot}$, where it is stated that the God who was known to the patriarchs as God Almighty (El Shaddai) now revealed to Moses His name

Jehovah, or Yahweh.[1] Elsewhere the name Yahweh is said to have been known and used in the pre-Mosaic period (e.g. Gen. 4^{26}, 15^7, 24^3). This seeming confusion is to be attributed to the presence in Genesis–Numbers of three main sources. The oldest, known as the Yahwistic source, or J, assumes that the name Yahweh was known in the pre-Mosaic period. The other two, the Elohistic source, or E (so called because of its use of the word 'Elohim', 'God'), and the Priestly source, or P, say that the name Yahweh was first revealed to Moses. Since in the Old Testament the name is not merely a convenient label, but an effective expression of the nature of the person named, the revelation of a new name of God represents a new beginning in religion. Accordingly, Exod. $3^{13\,f.}$ (E) and $6^{2\,f.}$ (P) are saying that such a new beginning was brought about through the work of Moses.

It might be supposed that, since Moses was brought up at the Egyptian court, the novel element in his teaching was Egyptian in origin. The heretic Pharaoh Ikhnaton was associated with a religious revolution which has often been regarded as monotheistic in character. But Israelite monotheism was very different from the faith of Ikhnaton; and it is a striking fact that outside the Wisdom books there is little evidence of Egyptian religious influence in the Old Testament. The biblical record itself points to a much more likely outside source of religious influence on Moses. His call came to him in the land of Midian, where he had married into a priestly family. It was there that the divine name was revealed to him. Later, just before the revelation of the law and the institution of the covenant at Sinai, Moses' father-in-law Jethro came to meet him, acknowledged the great deliverance wrought by Yahweh, and not only presided at a sacrificial act at which Aaron, the prototype of Israelite priesthood was present, but

[1] 'Yahweh' (or 'Jahveh') is generally accepted as representing the correct form of the word. In the course of time reverential motives led the Jews to avoid uttering this divine name. They replaced it by the word 'Adonai', 'Lord'. The absurd form 'Jehovah' arose from a mistaken transliteration of the consonants of Yahweh (Jahveh) and the vowels of Adonai.

gave advice on the administration of justice to Moses, Israel's supreme legislator. In Judges 1[16] it is stated that he was a Kenite (see R.V. margin). The Kenites, who may have had associations with the Midianites, appear elsewhere as zealous for Yahweh and friendly to Israel (Judges 4[11, 17 ff.]; 1 Sam. 15[6]). It may well be, therefore, that Moses was in some measure influenced from this quarter; but on the evidence at our disposal we cannot hope to define the nature and extent of such influence. What is clear is that the deliverance from Egypt left its stamp on Israelite religion.

Two general characteristics may be noted. First, Yahweh was a God who wrought mighty deeds in history. The awareness of the divine activity persisted through later developments of Israel's faith. The Exodus events were recalled. The prophets declared what Yahweh was doing and was about to do in the events of their own times; and this led to the thought of a final consummation as the outcome of God's purpose in history. Second, there was a prophetic element in the story. The Exodus did not happen simply as a series of awe-inspiring events which compelled faith. Moses, the man whom Yahweh had called and to whom Yahweh had revealed His purpose, was the herald and interpreter of the events. This anticipates the classical prophetic tradition in Israel; for it was an important part of the prophet's task to interpret events in terms of the will of God.

We must now consider (a) the nature of Yahweh, (b) His relationship with Israel, and (c) what He required of Israel, so far as these may be inferred from the Exodus traditions.

(a) From what has been said above it might be supposed that the name 'Yahweh' would provide an important clue to the nature of Israel's God. Various theories have been held about the etymology of the name. But these are largely irrelevant to our purpose. The question is not, 'What did the name originally mean?', but rather, 'What meaning did the name convey to the Israelites?' Probably an important clue is provided in Exodus 3[13 f.], where the name is associated with the verb 'to be': 'I AM WHO I AM' (or, as in R.S.V. footnote,

'I WILL BE, WHAT I WILL BE'). What is meant by the verb 'to be' is not bare existence, but existence manifested in activity. Yahweh reveals who He is by what He does; and this revelation will be continued in what He does for His people from generation to generation.

In the events of the Exodus Israel knew Yahweh as a Saviour God who had compassion on the afflicted slaves. The story also implies that He is Lord of the forces of nature, since He inflicted the plagues on Egypt, brought the Israelites across the Sea, and provided for them in the wilderness. Further, He does what He wills in Egypt, and is therefore not confined in His activity to the holy mountain, or, as might be supposed from some later passages (e.g. 1 Sam. 26^{19}; 2 Kings 5^{17}), to the land of Canaan. The gods of Egypt do not really enter into the story.

Is Yahweh, then, thought of at this stage as the one God? Was Moses a monotheist? The question involves an anachronistic use of the word 'monotheist'. Moses was not a monotheist in the later, abstract sense of the term. But there is justification for the claim which has sometimes been made that the Mosaic faith exemplifies a *practical* or *incipient* monotheism, i.e. that although there is at this stage no explicit denial of the existence of other gods, the germ of monotheism is already present. The two main factors which justify this claim are that in the Exodus traditions Yahweh is undisputed master of history and nature, doing as He will in every situation, and that He requires the undivided allegiance of His worshippers, displaying in this an intolerance which is hardly compatible with an admission of the effective existence of other gods. This practical character is evident even in the later stages of the Old Testament belief in Yahweh. It is never an abstract monotheism which is taught, but rather the fact of Yahweh's effective lordship in history and nature, and His right to undivided allegiance in national and individual life.

Yahweh's claim to exclusive allegiance is expressed in the statement that He is a *jealous* God (Exod. 20^{5}), an expression which should not be taken simply in the modern sense. It indicates Yahweh's active concern or *zeal* for His cause, and can denote negatively His intolerance of disloyalty and disobedience and positively His active concern for His people.

Yahweh is also a *holy* God. In their Old Testament use the terms 'holy' and 'holiness' refer first and foremost to the exalted majesty of Yahweh, to His otherness. He is 'the holy One'; and places, times, things, and people are holy only because of their relationship to Him. Holiness is not an abstract attribute of a remote deity. The holy God is the living, active God, who makes His presence known in the life of men. Nor should holiness simply be equated with morality or righteousness, though a close connexion between the two is recognized in some parts of the Old Testament.

(*b*) Israel is called to be 'a holy nation' (Exod. 19[6]). This means that Israel belongs to Yahweh. Before the organization and structure of a nation-state were worked out under the monarchy, the confederacy of Israelite tribes had a unity derived from their common relationship to Yahweh. Two thoughts are important here.

First, they owed their unity to what Yahweh had done for them. In some parts of the Old Testament this is expressed by saying that Yahweh *chose* Israel; they were in that sense the *elect* people, owing their existence not to their own achievements but to the action and purpose of Yahweh.

Second, this relationship was expressed in a covenant. The covenant idea is of fundamental importance in Old Testament religion. It expresses a relationship in which the element of obligation is present. In the stories of the covenants with Noah (Gen. 9[8 ff.]) and Abraham (Gen. 15, 17) God promises that He will do certain things; and accordingly the emphasis is on the divine pledge. But in the story of the covenant which follows the Exodus what is made explicit is the obligation which will rest on Israel once the covenant has been established. Yahweh's care for His people, as His side of the covenant, is, of course, also involved. He has already delivered them; and among the results of the deliverance will be the gift of the promised land and the blessings which He will give them there; but the most important outcome of the deliverance is the relationship with Israel which is established in the covenant. The essence of this covenant is tersely summed up in the formula, 'I will take you for my people, and I will be your God' (Exod. 6⁷; cf. Jer. 31³³).

It has sometimes been supposed that the covenant idea represents a later theological element which has been superimposed on the narratives. But the covenant forms, as represented by the Ten Commandments (Exod. 20¹⁻¹⁷), is strikingly parallel in structure to treaties made between Hittite kings and their vassals in the period 1450–1200 B.C. The vassals are reminded of what the king has done for them and of their obligations to him: allegiance, tribute, service, acceptance of his jurisdiction, and the like. So Israel is reminded of Yahweh's mighty acts, and summoned to respond in loyalty and obedience.

From this it follows that the Israelite community was constituted not by ties of blood but by Yahweh's act. They belonged to each other because He had made them His own. A further consequence is that the obedience which is required of Israel is the grateful response to what Yahweh has done. Nothing could be further from the truth than the notion that Israel boxed God up in a set of commandments. The Ten Commandments appropriately begin, 'I am the LORD your God, who brought you out of the land of Egypt, out of the house of bondage' (Exod. 20²). The gracious act of the Saviour God is the presupposition of the commands laid on Israel.

(c) To determine what were the specific requirements made of Israel in the Mosaic age is far from easy. The mass of legislation contained in the Pentateuch is a complex literary compilation. The whole of Leviticus and much of the legislation in Exodus and Numbers comes from the Priestly Code, which, though it may embody much ancient material, was probably compiled during or soon after the Exile. In it is embedded the Code of Holiness (Lev. 17–26), which is of uncertain date and provenance. The Deuteronomic Code (Deut. 12–26) is in all probability the book which was found in the Temple in the reign of King Josiah (2 Kings 22): some features in it reflect seventh-century conditions; but its staple material must be older. Exod. 20^{22}–23^{33}, the Book of the Covenant, presupposes settled agricultural life. It has interesting resemblances to the Code of Hammurabi, and also to Assyrian, Hittite, and Sumerian law codes; but no doubt such elements were mediated to Israel through Canaanite influence. This collection combines in an interesting way two types of law: *casuistic* law, which is hypothetical in form (e.g. Exod. 22^{1-17}) and has its appropriate setting in the courts, where cases would be argued, and *apodictic* or categorical law, expressed in terms of 'you shall' and 'you shall not' (e.g. Exod. $22^{21 \text{ f.}}$), which has its appropriate setting in the religious assembly. But the whole collection is regarded as expressing the will of Israel's God.

Of special interest for our present purpose are the familiar Ten Commandments in their two forms (Exod. 20^{1-17}; Deut. 5^{6-21}) and the short code in Exod. 34^{12-26}. The latter is sometimes called the 'Ritual Decalogue'. A little arithmetical adjustment is needed to make the number of commandments which it contains total no more than ten; but its ritual interest is obvious. By contrast, the term 'Ethical Decalogue' is applied to Exod. 20^{1-17}; Deut. 5^{6-21}. This is not entirely apt, since the commandment about the Sabbath, for example, is a ritual commandment; but the presence of a strong ethical emphasis cannot be denied. It has sometimes been argued as a general principle that the ethical interest in religion is a later development than the ritual interest, and, in particular, that the Ethical Decalogue reflects and is derived from the moral teaching of the eighth-century prophets. But the supposed general principle is questionable; and it is arguable that the moral teaching of the prophets is based on traditional standards of which the Ethical Decalogue is an ancient formulation. In substance the Ten Commandments are so representative of Israel's religious and moral precepts, that it is difficult to think of a period after the conquest at which they would have represented an innovation. It is not unreasonable to suppose that, possibly in a terser form, they are of Mosaic origin. The

Stele of Hammurabi.

interesting suggestion has been made that the Ritual Decalogue, which belongs to the southern source, J, was a code
which was adopted by a group of tribes who moved into
Canaan from the south and who acquired their knowledge of
the worship of Yahweh from the Kenites by a process of gradual
absorption, whereas the Ten Commandments are in essence
the code given to those tribes who crossed the wilderness under
Moses and invaded Canaan under Joshua, tribes whose
experience of divine deliverance and of the need for decision is
reflected in the ethical character of their code. However that
may be, the familiar Ten Commandments are an effective
expression of fundamental religious and moral standards in
ancient Israel and of their relation to Yahweh's saving acts.

Of the sacred objects associated with the wilderness period
the most interesting and important is the Ark. This was a chest,
made of acacia wood, which was carried by the Israelites in
their wilderness wanderings. After the conquest it was kept for
a time at the sanctuary at Shiloh, was captured in battle by the
Philistines, came to Kiriath-jearim, and was brought up by
David to Jerusalem, where it finally found a place in the Holy
of Holies in Solomon's Temple. If we ask what was the religious
significance of an object which had so long a history, we must
reckon with the possibility that it was understood in different
ways at different times. The story of its being brought into the
camp to ensure victory (1 Sam. 4; cf. 2 Sam. 11[11]) suggests
that it represented the mighty presence of Yahweh with His
people particularly in what were regarded as holy wars
against the enemies of Yahweh and His people. This is in
keeping with the ancient formulas in Num. 10[35 f.], where the
Ark is addressed in these terms:

> 'Arise, O LORD, and let thy enemies be scattered;
>   and let them that hate thee flee before thee.'
> 'Return, O LORD, to the ten thousand thousands of Israel.'

Later it came to be thought of as the throne of the invisible
Yahweh. In Deut. 10[5], it is the receptacle for the tables of the
law, hence no doubt the titles, 'the Ark of the covenant', and

'the Ark of the testimony'. It may have been connected in some way not only with the revealed will of Yahweh but with the revelation of His will (1 Sam. 14[18]). But as the potent symbol of the presence of Yahweh, the palladium of His marching host, it clearly is appropriate to the wilderness period and to a religion in which image worship was forbidden.

According to Exod. 33[7-11] Moses resorted to a tent ('the tent of meeting') outside the camp in order to commune with God and learn His will. This differs in location and character from the tabernacle described in great detail in the Priestly source (Exod. 25-27, 30, 36-40) which was in the centre of the camp, and which housed the Ark and other items of sacred furniture. This complicated structure has sometimes been regarded as a projection back into the Mosaic age of the main features of Solomon's Temple. But in David's reign the Ark was housed in a tent in Jerusalem (2 Sam. 7[2]); and there may be some connexion between this structure and the accounts of the Tabernacle. But however great the uncertainty about details, it is clear that there was in Israel a long-standing tradition of a sacred tent which was obviously appropriate to the conditions of the wilderness and which survived into the period after the conquest.

Although the prophets Amos (5[25]) and Jeremiah (7[22]) seem to deny that sacrifice took place or was commanded in the Mosaic period, there can be no doubt that sacrifices of some kind were carried out then (see, e.g., Exod. 18[12]; 24[5-8]), though they would necessarily be considerably simpler than in later times.

Of the festivals which the law prescribes, the Passover is the one most intimately associated in Israelite tradition with the Exodus, and therefore with the Mosaic stage of Israel's religion (Exod. 12; especially vv. 21-27). A lamb was killed; its blood was splashed on the door-posts and lintel; and its flesh provided a sacrificial meal. It was the beginning of a period of seven days during which unleavened bread was eaten. But it is commonly held that the feast of Unleavened Bread was one of three agricultural festivals which Israel adopted after settling in

Canaan (the others being the feasts of Weeks and of Ingathering or Tabernacles), and that in the course of time the Passover (a pastoral rite) and the feast of Unleavened Bread (an agricultural rite) were combined because they were celebrated at the same time of the year. But the history of the festival presents complex problems; and since semi-nomads, such as the Hebrews were before settling in Canaan, would have some experience of growing cereal crops, a ritual connected with the new cereal crop would not necessarily be foreign to their way of life.

The observance of the Sabbath as a weekly day of rest came to have great importance in later ages when it was one of the more obvious features which marked the Jews off from their non-Jewish neighbours. The name is connected with a verb which means 'to desist'. In the creation story with which the book of Genesis begins, the observance of the seventh day as a day of rest is related to God's resting, or desisting from His work, after the six days of creation (Gen. 2^{1-3}). This is the reason given in Exod. 20^{11} for keeping the Sabbath. In Deut. 5^{15}, on the other hand, it is connected with the release from hard bondage which the Exodus brought. In the Ritual Decalogue (Exod. 34^{21}) and in the Book of the Covenant (Exod. 23^{12}) there is a command to refrain from work on the seventh day; but the name 'Sabbath' is not used, though the corresponding verb is. In some passages in the historical and prophetical books the Sabbath is coupled with the new moon (2 Kings 4^{23}; Isa. 1^{13}; Hos. 2^{11}; Amos 8^5). The question arises whether, in the pre-exilic period, the Sabbath, like the new moon, was a monthly and not a weekly observance. It is suggested that in the Decalogue the original form of the Sabbath commandment was simply 'Remember the sabbath day, to keep it holy' (Exod. 20^8), or 'Observe the sabbath day, to keep it holy' (Deut. 5^{12}), and that the reference which follows to the seventh day was a later addition, dating from a time when the Jews, under Babylonian influence, had changed the Sabbath into a weekly observance. But the case for Babylonian influence is not strong. In Babylonia in at least two months certain actions had to be avoided on the seventh, fourteenth, twenty-first, and twenty-eighth days; but these, unlike the Sabbath, were days of evil omen. The fifteenth day of the month was called *shapattu*; but it was not a rest day. No doubt we must recognize that the Sabbath, like other religious institutions, had a history; but it is not unreasonable to suppose that the weekly day of rest which is referred to in Exod. 23^{12}, 34^{21} is identical with the Sabbath which we know to have been observed after the Exile and with the Sabbath of the Decalogue, and that this institution goes back to Mosaic times.

As it now stands, the Pentateuchal legislation is complex and represents a long process of development; but the dominating thought underlying the whole is that observance of these laws is part of Israel's response of gratitude to the God who brought her out of bondage in Egypt.

IV

FROM THE SETTLEMENT TO SOLOMON

Adjustment and Consolidation: the Age of the Judges

IT is notoriously difficult to draw clear lines of historical de-marcation separating one period from another, not least when there are serious gaps in the evidence. But from the end of Joshua's life to the beginning of Saul's reign there is a chapter in Israel's story which is recognizably different from what preceded it and from what followed. The necessary sequel of the conquest was a period of adjustment and consolidation. The semi-nomad had to become a farmer, and in the process was exposed to the influence of Canaanite culture. In different parts of the country there were political pressures and military dangers which had to be faced. As time passed and new factors were brought to bear on the life of the confederacy, the need for a stronger system of centralized organization became pressing. In every situation the religious element was present in some way, affecting, and being affected by, all the changes that came about.

Here, as in other periods between the conquest and the Exile, we are above all dependent on the Deuteronomistic history (see above, p. 5), and must remember that it is not, as it appears in our Bible to be, a sequence of separate books, but one history, based indeed on earlier material of various kinds, but representing a particular interpretation of Israel's life, with a strong emphasis on the ideal unity of all Israel and an uncompromising hostility to Canaanite influence.

In Josh. 24 there is a description of an assembly of the Israelite tribes at Shechem at which Joshua reminded the tribal leaders of what Yahweh had done for them and sum-moned them to give their allegiance to Him and enter into covenant with Him. This represents the culmination of the process of conquest and of the allocation of the land to the tribes; and, on the face of it, it is a renewal of the covenant made with a previous generation in the wilderness. But it may well involve something more. We have noted the possibility that there were Hebrews who did not go down to Egypt but remained in Palestine. For them the covenant at Shechem would be the means of entering into the life of the confederacy and sharing in all that was summed up in the confession of Yahweh's mighty acts. Since the covenant community was the outcome not of racial factors but of the divine deliverance, these and other elements could enter into the covenant bond. Canaanites and Amorites, Hittites, Girgashites, Perizzites, and Horites had inhabited the land, a varied population, partly Semitic, partly non-Semitic, displaying no wide political unity except such as was imposed on them from outside. That some of them came to form part of Israel is entirely probable. Shechem seems to have been an important centre of the fusion of Israel with the earlier inhabitants of the land (cf. Judges 9).

The word which is currently used to designate the Israelite confederacy is 'amphictyony'. The term is borrowed from Greek usage and denotes an association of communities or tribes centring in an ancient shrine at which they met periodically. The parallel should not be pressed; for there were special factors in the Israelite situation. It does appear that the presence of the Ark at a sanctuary in Israel made that sanctuary the central point for the worship of the tribes (see, e.g., the references to Shiloh in 1 Sam. 1–3). But the resting place of the Ark was not always the same; and accordingly the central sanctuary was located in different places at different times. Further, although there is fairly early evidence for the list of twelve tribes (e.g. Gen. 49), it is probably a mistake to think of it in rigid terms, particularly at this period. Reuben and Simeon waned considerably in importance. Ephraim and Manasseh seem to have developed out of an original tribe of Joseph. For a considerable time the tribe of Judah seems to have been in a somewhat detached position. The absence of any reference to it in the Song of Deborah (Judges 5) is significant.

During the period after the conquest Israel's security and independence were threatened in two main ways. On the one hand, some of the tribes were temporarily subdued by their Canaanite neighbours or by settled peoples beyond the borders of the land. On the other hand, parts of the country were harried by raiders from the east. These situations are described in the central part of the book of Judges (2^6–16^{31}).

The first story is a puzzling one. Othniel, whose home was in the south (Judges 1^{11-15}) is said to have delivered Israel from Cushan-rishathaim, king of Mesopotamia (Judges 3^{7-11}). The oppressor's name (Cushan of double wickedness) seems artificial; and his identity has not so far been established. It has been suggested (assuming a slip which is much more easily made in Hebrew than in English) that he ruled not in Mesopotamia but in Edom.

The next oppressor to be mentioned is the Moabite king, Eglon, whose tyranny was ended by the resourcefulness and cunning of the Benjamite, Ehud. The assassination was followed by a revolt which broke Moabite power.

Two accounts have been preserved of a victory over the Canaanites won by tribes in the north: a prose version in Judges 4, and in Judges 5 the magnificent ancient poem, the Song of Deborah, which is probably contemporary, or nearly so, with the events. Inspired by the prophetess Deborah, Barak, a warrior of the tribe of Naphtali, assembled the forces of the tribes at Mount Tabor and led them against the Canaanite army under Sisera. The engagement is of considerable interest for two reasons. It is the only instance on record of a major conflict between Israelites and Canaanites after Joshua's campaigns; and it was won over a powerful force of chariots operating on the plain. The early victories of the Israelite invaders seem to have owed much to the element of surprise, to guerrilla tactics, and to the skilful use of terrain. They were at their best in the hills. On the plains they were at the mercy of the chariots with which their opponents were equipped (Josh.

$17^{16, \, 18}$; Judges 1^{19}). But on this occasion the chariots were immobilized by a storm which turned the plain into a sea of mud.

The scene now changes to the central hill country. The Israelite farmers were being pillaged repeatedly by Midianites, Amalekites, and other eastern tribesmen. A Manassite called Gideon was called by Yahweh to lead a counter-attack, which he did with a force which had been deliberately and drastically reduced in numbers but with such success that centuries after-wards 'the day of Midian' was still used as an apt description of a decisive victory (Isa. 9^4).

Although the story includes an account of an attack on Canaanite Baal-worship (Judges 6^{25-32}), there can be little doubt that the common need to repel raiders was one of the factors which helped forward the fusion of Israelites and Canaanites; and it is interesting to note that Gideon's extensive harem included at least one Canaanite woman (Judges $8^{30 \, f.}$), whose son, Abimelech, exercised a local kingship in Shechem (Judges 9). It is said that Gideon himself was offered the status of king but refused it (Judges $8^{22 \, f.}$).

The aggression of the Transjordanian kingdom of Ammon against Gilead was repelled by Jephthah, a Gileadite outlaw (Judges $10^{17}-12^7$). His victory established his personal authority, but also led, because of his own rash vow, to the sacrifice of his daughter. This profoundly tragic incident is of great religious interest, both as showing that human sacrifice was sometimes practised in Israel, and also because of the reference to an annual ritual of lamentation, which probably had connexions with a fertility cult.

In the stories about Samson (Judges 13–16) we find not only a somewhat different kind of deliverer but also an essentially different kind of danger. Samson appears to the modern reader to have been a combination of the strong man and the buffoon. As a Nazirite (Judges $13^{5, \, 7}$, 16^{17}), he was consecrated to Yahweh in a special way; but it is clear that he did not observe all the restrictions which are elsewhere said to have been applied to Nazirites (Num. 6^{1-21}). We must, of course, remember that consecration to the service of a deity did not necessarily have the moral implications which in modern times would be thought appropriate. The essential thing was not 'Puritanism' in the popular modern sense, but the fact of being set apart to the service of the deity. Unlike the other Judges, however, Samson did not act as an inspired and inspiring leader. His exploits were carried out single-handed; they did not involve the rallying of his own or other tribesmen. He was a Danite; and the stories about him are set in the Shephelah region, which the tribe of Dan occupied before it moved northwards (Judges 18).

The reason for the migration was Philistine pressure. The Philistines formed part of the great movement of peoples in the lands around the eastern Mediterranean which followed the collapse of the Minoan empire. Their name occurs in Egyptian inscriptions of the thirteenth century and earlier, denoting one element in a group called 'the Sea Peoples'. Their armour, and, in particular, their distinctive feathered head-dress, are familiar from Egyptian reliefs. Their attempts to invade Egypt were repelled; and various elements settled along the coast of Palestine, the Philistines occupying the stretch from Joppa to Gaza, where their five leading cities were Ekron, Ashdod, Ashkelon, Gath, and Gaza. They had a higher level of material culture than the Israelites, and, in particular, were masters of the craft of producing iron tools and implements. They were a much more serious threat to Israel than any other hostile power at this period; and they came near to dominating the entire country. Samson's enterprises against them were no doubt carried out at a time when the power of the Philistines was beginning to be developed. Later the magnitude of the danger became clearer; and in the time of Samuel's mentor, the priest Eli, and of Samuel himself, it seemed that the Philistines might finally subdue Israel. Just before the end of Eli's life they won a double victory over the Israelites (1 Sam. 4) and captured the Ark. From the references to their later activities it is clear that they were able to establish their hold on widely separated parts of Israel.

Prisoners from among the Sea Peoples.

The stories of the crises and deliverances which are described in the book of Judges make, as they stand, a connected narrative. But important differences are evident between the stories themselves and the connecting framework which surrounds them. The framework has three significant characteristics: (a) a chronological scheme which is part of a larger system linking the Exodus with the founding of the Temple (1 Kings 6[1]; cf. above, p. 25); (b) a series of religious comments, in which the crises are represented as Yahweh's punishment on Israel for apostasy, and the deliverances as His response to the people's appeal for help; (c) the view that the crises and deliverances affected all Israel and that the Judges who led the people to victory exercised authority over the entire country. But if the stories are read separately from the framework, the impression they give is that each of them refers to a particular area of the country and to a particular tribe or group of tribes, from which it follows that the leaders mentioned had local rather than national authority. The stories are the material which the Deuteronomistic historian used; the framework is the means which he employed to arrange and interpret that material. He has pointed the characteristically Deuteronomic moral of the sequence of apostasy and affliction, or repentance and deliverance; and he has emphasized the characteristically Deuteronomic ideal of the unity of all Israel. But if the crises did not affect all Israel but were regional, and if the Judges exercised only a local and not a national authority, then the episodes need not have happened in sequence, but may have overlapped in time; and the period of the Judges may well have been considerably shorter than might be supposed from the figures given throughout the book.

A distinction is generally made between the major and the minor Judges. Only brief accounts are given of the minor Judges; Shamgar (Judges 3[31]), Tola (10[1f.]), Jair (10[3-5]), Ibzan (12[8-10]), Elon (12[11f.]), and Abdon (12[13-15]). It is the major Judges (Othniel, Ehud, Deborah, Gideon, Jephthah, and Samson) who appear in the stories of crisis. Their prime task is to be the human agents in Yahweh's deliverance of His people.

Within that context and in that sense they are saviours. Of most of them it is said that the spirit of Yahweh came upon them. Apparently there was a sudden accession of power, which was demonstrated in abnormal qualities of courage, skill, strength, and the gift of leadership. It was because of these *charismatic* gifts (i.e. gifts bestowed by God's grace) that the Judges came to exercise authority, and not in virtue of any formal appointment or lineal succession. The story of the offer of the kingship to Gideon (Judges 8²²ᶠ·) indicates the possibility of a transition from the purely charismatic leadership to an inherited monarchy. This possibility became a reality when the monarchy was fully established, though the tradition of charismatic leadership was by no means superseded. During the period of the Judges the organization of the tribes remained somewhat loose; and it was no doubt the succession of crises and the exercise of authority by one man, though that authority was personal, local, and temporary, which prepared the way for the monarchy. The final stimulus was given by the Philistine menace.

The Rise of the Monarchy: Saul

Samuel is represented as the first and greatest kingmaker in Israel. It was through him that the monarchy was introduced into Israel and by him that both Saul and David were anointed. He had been brought up at the sanctuary at Shiloh under the guidance of Eli, the priest, in whose time Philistine pressure on Israel was severe. Israel was twice defeated at Aphek and the Ark was captured (1 Sam. 4). Archaeological evidence indicates that Shiloh was destroyed, though there is no reference to this event in the narrative in 1 Samuel (cf. Ps. 78⁶⁰; Jer. 7¹⁴). The Philistines strengthened their hold on the land by establishing garrisons at key points. In Samuel's time they continued to keep Israel in subjection, making resistance more difficult by the measures they took to retain a monopoly in iron (1 Sam. 13¹⁹⁻²²).

Samuel is said to have judged Israel (1 Sam. 7¹⁷); but the application of this verb to his activity seems to indicate not a

sudden access of divine power leading to a dramatic deliverance but the exercise of some kind of administrative and
judicial authority (1 Sam. 7$^{15-17}$) comparable in character to
that of the minor Judges. There is, indeed, a record of an
important victory won by Israel over the Philistines during the
judgeship of Samuel (1 Sam. 7$^{3-12}$); but the account of the
establishment of the monarchy shows that Philistine domination continued.

In the chapters which tell of Saul's elevation to the kingship
(1 Sam. 8–12) at least two strands are interwoven. One
(1 Sam. 9^1–10^{16}) presupposes Philistine oppression and represents the new leader as Yahweh's instrument for the deliverance
of His people. The identity of the prospective king was divinely
revealed to Samuel, who privately anointed him. Saul's
charismatic endowment was demonstrated in his leadership
of the Israelite forces against the Ammonites who had be-
leaguered the town of Jabesh-Gilead, east of Jordan (1 Sam.
11). The other strand (1 Sam. 8, 10$^{17-27}$, 12) presents the
request for a king as a rejection of the authority of Yahweh and
is critical of the monarchy. It tells how the king was chosen by
means of the sacred lot at a national assembly convened by
Samuel, who is here much more obviously a national figure
than in the other account. Though the former of these strands
is probably the nearer in time to the events, it is a mistake to
overemphasize the differences between the two presentations
and to assume that the later one is simply unhistorical. Not
only individual kings, but also kingship as such seems to have
come under criticism in certain circles in Israel; and at the
time of its establishment it may well have been regarded by
some as a dangerous innovation, however important the
benefits which it conferred.

The benefits were not easily secured. Although Saul's first
achievement was the defeat of the Ammonites, he had throughout his reign to face the far more serious threat of Philistine
domination. Clearly some headway was made in the task of
uniting and liberating Israel. An enterprising exploit by
Jonathan and his armour-bearer at Michmash led to a victory

of some importance and the rallying of some hitherto demoralized Hebrews to Saul's army (1 Sam. 14). David's personal triumph over a huge Philistine champion made possible another victory, this time, significantly, not in the heart of Israelite territory but on the approaches to Philistia (1 Sam. 17). The references to David's later activities in Saul's service show that Israel was by no means always on the defensive. But in the end the Philistines decisively defeated Saul at the battle of Mount Gilboa in the plain of Esdraelon (1 Sam. 31).

Throughout much of Saul's reign three other conflicts are evident. At an early stage there was a breach between him and Samuel. Two incidents are recorded. In the first (1 Sam. 13$^{8-14}$), when Samuel failed to arrive on time Saul himself undertook the sacrifices which inaugurated a campaign against the Philistines. The second (1 Sam. 15) concerns a campaign against the Amalekites in the far south of the country. Saul was instructed to take neither prisoners nor spoil. Such total destruction of the enemy and his goods was known as *ḥērem* and was a practice by which the fruits of victory were made over to the deity (cf. Num. 21$^{2-4}$; Deut. 20$^{10-18}$; Joshua 6$^{17, 21}$). Because Saul did not apply the *ḥērem* strictly, he was denounced by Samuel and told that he had been rejected as king. His reign did not in fact end there and then; but clearly his authority and confidence must have been impaired by the knowledge that the prophet had solemnly declared Yahweh's displeasure with him.

The second conflict may in part have arisen from the first. After his anointing by Samuel, the spirit of Yahweh came upon Saul, imparting to him the prophetic frenzy and empowering him to lead Israel to victory. Now he became subject to moods of gloom and violence. It was as if the abnormal energy by which his early successes had been gained had now been turned against the king himself.

According to 1 Sam. 16$^{14-23}$ (contrast 17$^{55-58}$), David the Bethlehemite was added to Saul's retinue so that, by his skill as a musician, he might bring relief to the king in his times of mental stress. But it is evident that jealousy of David was yet another factor which marred Saul's later career. The chief reasons for this were no doubt David's success as a warrior and his popularity, the close bond between David and Saul's son Jonathan, and the fact that David was an obvious candidate for kingship. Much of the record of Saul's reign is concerned with the worsening relations between the two men: first Saul's treacherous malevolence, then David's outlawry in the hill country of Judah, and finally his exile in Philistine territory, so that when Saul's last struggle with the Philistines took place David and his men were not in the Israelite ranks.

Saul's reign began with a successful military action which is reminiscent of the exploits of the Judges; and in many ways he resembles these earlier leaders more than the later kings. He gathered around him a force of warriors who formed at least the nucleus of a standing army; but there is no indication that he developed anything like the royal household and administrative service which existed in later times. He reigned in simple state at Gibeah. From the record of his campaigns, however, it is evident that he exercised an effective authority over at least the southern and central parts of the country; and his relief of Jabesh-Gilead won lasting loyalty for him and his house east of Jordan. But such national unity as he had established was seriously impaired by his relations with David. What he had achieved had in large measure been undone when his reign ended.

The Reign of David

Saul's death left Israel at the mercy of the Philistines and internally divided. David, who had been in the somewhat questionable position of a vassal of Achish, king of Gath, now moved to Hebron, where he was anointed king of Judah (2 Sam. 2[1-4]). But at Mahanaim east of Jordan, Saul's son Eshbaal[1] was established as king. The real power in his

---

[1] This form of the name, which is found in 1 Chron. 8[33], 9[39], is to be preferred to 'Ishbosheth' (2 Sam. 2[8], etc.), in which 'bosheth' (shame) has been deliberately substituted for 'baal', for religious reasons.

attenuated kingdom seems to have been Abner, Saul's commander-in-chief (2 Sam. $2^{8f.}$).

David's advance to authority over all Israel was facilitated by a grim sequence of personal and family feuds. After an encounter between a force of David's men under Joab and a similar party led by Abner, Asahel, Joab's brother, pursued Abner but was killed by him. This created a blood feud between Abner and Asahel's brothers, Joab and Abishai (2 Sam. 2). Later, Abner quarrelled with Eshbaal and offered David his support, promising to secure for him dominion over all Israel. On hearing of the agreement, Joab hurried after Abner and treacherously murdered him, thus removing the one really strong leader of the central tribes (2 Sam. 3). Some time afterwards Eshbaal was assassinated (2 Sam. 4); and the tribes which until then had retained their allegiance to the house of Saul now offered the kingship to David. In a remarkable way the field had been cleared for him by the violence committed by others, from which he had publicly dissociated himself.

The extension of David's royal authority over Israel was established by a covenant (2 Sam. 5^3). It has been claimed that no real unity was established between Judah and the central and northern tribes other than that which was maintained by David's personal authority. Later events such as Absalom's revolt, and, even more decisively, the disruption of the kingdom, show that serious tensions remained. But at all events the unity seems to have been sufficiently effective to rouse the Philistines to action. David had been a vassal of the Philistines. As king of Judah alone he might perhaps be disregarded. As king of Judah and Israel he was a serious threat to Philistine power. They moved against him, but were twice defeated (2 Sam. 5^{17-25}). Such unity as had been achieved was reinforced by the winning of independence from foreign domination.

It seems likely that these victories preceded the capture of Jerusalem (2 Sam. 5^{4-10}), though they follow it in the narrative. Not only was it an outstanding military feat, but it marked a further step in the consolidation of David's royal authority and the unification of his kingdom. It provided him with a centre from which he could exercise authority over the country far more effectively than from Hebron in the remote south. Jerusalem lay between the territory of Judah and that of the central and northern tribes, and it had belonged to neither. The choice of this city as the new capital was a shrewd move. David further enhanced its prestige by bringing the Ark from Kiriath-jearim and installing it in a sacred tent at Jerusalem (2 Sam. 6). In this way he established a powerful link between the hitherto non-Israelite city and Israel's religious traditions and so laid the foundation of the later religious dominance of Jerusalem.

By a succession of victorious campaigns (2 Sam. 8, 10–11) David further strengthened the position of his kingdom, providing it with a semicircle of buffer territories and tributary states and gaining control over several lucrative trade routes. The Philistines had already been quelled. East of the Dead Sea, the Moabites were subdued. A campaign against Ammon proved difficult, partly because of the help given to the Ammonites by the Aramaeans in the north. Only when the Arammaeans had been defeated, and after a costly siege, was the Ammonite capital taken. Victory over the Edomites was followed by ruthless measures against the royal house and the population, and by the establishment of a provincial administration. Control of this territory gave access to the Red Sea, and also to important mineral resources. Further operations were undertaken against the Aramaeans; and as a result David's sway, and the area from which he received tribute, were extended to Damascus and beyond. Also in the north, but further west, amicable relations were established with Phoenicia (2 Sam. 5[11]), a link which was to prove important in Solomon's reign and in the later history of Israel.

Thus, at a time when there was no great power to intervene, David established Israel as an independent state with extensive dependencies, with access to important natural resources, and with influential commercial connexions. Within a single generation there was a striking advance in the general standard of living, a development which was impressively continued in the next reign. A royal administrative system began to be organized on Egyptian lines; and in the military field a standing force of foreign mercenaries (the Cherethites and the Pelethites; 2 Sam. 8[18], 20[23]) was maintained, in addition to the Israelite militia (the army; 2 Sam. 8[16], 20[23]). The king also had a personal bodyguard. David's royal household was considerably more pretentious than that of Saul, though modest as compared with that of his successor.

Though David had conspicuous faults, he appears to have exercised a remarkable personal influence and to have retained men's loyalty in spite of his errors and failings. No doubt his magnetic attractiveness as well as his military prowess helped the new state through the strains and tensions which were inseparable from its rapid development. Two problems had to be faced: the maintenance of the somewhat precarious unity of the state and the choice of a strong successor. If Saul's daughter Michal had borne David a son, perhaps both problems would have been more easily solved. Amnon, his eldest son by another wife, was murdered by Absalom because of the wrong done to Absalom's sister Tamar. After going into exile because of David's displeasure, Absalom was permitted to return. He exploited national dissension and grievances against David's administration to win support for himself, contrived to get himself anointed king in Hebron, and marched on Jerusalem. Accompanied by most of his personal retinue, David left the city and escaped east of Jordan, where shortly afterwards Absalom's army was routed and he himself killed by Joab. But even in the restoration of David to the capital further dissension broke out between Judah and Israel, which led to a revolt headed by a Benjamite named Sheba. It was promptly subdued; but it was one more sign of the internal strains in David's kingdom.

These events are recorded in one of the most remarkable narratives in the Old Testament, which is probably almost contemporary with David's reign. A masterpiece of classical

Hebrew prose, it portrays with profound psychological insight the interplay of personalities and events at David's court as the problem of the succession to the throne moves towards a solution. The main part of the story is contained in 2 Sam. 9–20; but the climax is reached in 1 Kings 1–2. With Amnon and Absalom out of the way, another of David's sons, Adonijah, tried to make good his own claim to the throne. The king was old and incapable of personal intervention or initiative. Adonijah secured the support of Joab and of Abiathar the priest. Both of these men had been with David in his early days of hardship and danger; and it may be that they represented a conservative party at the court. They were opposed by another group, headed by Benaiah the commander of the mercenaries, Nathan the prophet, and Zadok the priest, who sponsored Solomon, the son of David's favourite wife Bath-sheba. David had seduced her and engineered the death of her first husband Uriah; and for this double crime he had been bluntly denounced by Nathan (2 Sam. 11–12). Now, in the final crisis of David's reign, we find the court prophet helping Bath-sheba's son to the throne. With the aged king's sanction Solomon was anointed and publicly acclaimed king. Adonijah and his associates had been outwitted. Not only a new king but a new régime had been established.

The Reign of Solomon

The *coup d'état* by which Solomon gained the throne was followed, shortly after David's death, by the brutal removal of possible rivals and opponents. Adonijah, Joab, and Shimei (a member of Saul's family) were killed; and Abiathar the priest was expelled from the capital to his patrimony in Anathoth. Although we are told (1 Kings 2¹⁻⁹) that David had on his deathbed instructed Solomon to carry out two of these acts, they give an unpleasant indication of what seems to have been one element in Solomon's reign, the ruthless exercise of royal power. In tradition he has been celebrated as the father of Israelite Wisdom, the builder of the Temple, the king who, after the almost unbroken warfare of his father's reign, gave

Israel a period of peace and of fabulous prosperity. But it is clear that there was another and a harsher side to the story.

By contrast with the account of David's reign, there is an almost total absence of dramatic movement in the record of the achievements of Solomon (1 Kings 3–11). The Deuteronomistic historian has assembled lists, statistics, annalistic fragments, and stories of various kinds round what was for him Solomon's supreme undertaking, the building of the Temple; but he has not given a chronologically consecutive narrative. It is not possible to trace the progress of events; but the chief aspects of Solomon's policy are sufficiently clear.

The Temple, Solomon's most conspicuous legacy to later ages, is a significant pointer to some of these aspects: the combination of opulence and extravagance, the exacting demands made on the nation's resources, the important foreign contacts. It was situated just north of David's city, and was orientated east and west. A porch led to the main hall (the Holy Place), behind which lay the dark inner shrine (the Holy of Holies) where the Ark rested. The general plan was derived from Phoenician and Canaanite temples. Phoenicia supplied both materials and skilled craftsmen for the work. It has often been described as a royal chapel. It was something more, for the Ark gave it the status of Israel's national sanctuary. But it did form part of a complex of royal buildings (1 Kings 7$^{1-12}$) and thus was closely linked with the royal house and household.

This ostentatious building programme was matched by the maintenance of an extensive royal establishment of courtiers and officials and a large harem. Some impression of its size and of the corresponding drain on the national resources is given by the account of its daily requirements (1 Kings 4$^{22-23}$).

So that the resources of the country might be more effectively exploited, Solomon divided Israel into twelve administrative districts (1 Kings 4$^{7-19}$). Not only did the people have to sustain a heavy burden of taxation in kind; forced labour was also imposed on them. According to 1 Kings 9$^{20-22}$, only the non-Israelite population had to serve in this way; but 1 Kings 5$^{13-14}$ states that 'all Israel' was affected, which is corroborated

The Temple Area in Jerusalem.

Architectural reconstruction of Ezion-geber.

by the complaint made by the people at the beginning of the reign of Solomon's successor (1 Kings 12$^{1-20}$).

Solomon was overspending the wealth of the country. One indication of his financial embarrassment is the transfer of some Israelite towns to Tyre (1 Kings 9$^{10-14}$). On the other hand, his reign was undoubtedly one of intense commercial and industrial activity and of expanding trade, in which Hiram, King of Tyre, was an active partner. David's conquest of Edom had given Israel access to extensive deposits of copper. Modern archaeological investigation has shown that the metal was mined, smelted, and refined on a large scale. A substantial refinery has been excavated on the site of Ezion-geber, which was also, by reason of its situation at the head of the Gulf of Akaba, a base from which Solomon's merchant fleet, manned by Phoenician sailors, voyaged on trading missions along the coasts of the Red Sea and beyond (1 Kings 9$^{26-28}$, 10^{11} f.). Solomon's commercial interests in that region indicate the

Stables at Megiddo.

reason for the visit to Jerusalem of the Queen of Sheba (1 Kings 10¹⁻¹⁰, ¹³), whose kingdom lay in southern Arabia. Another of Solomon's enterprises was the trade which he carried on as a middleman in horses and chariots bought from Asia Minor and Egypt (1 Kings 10²⁸ᶠ·).

Solomon further strengthened the position of his country by political and matrimonial alliances. His marriage to an Egyptian princess is recorded as of special importance (1 Kings 3¹, 7⁸, 9¹⁶). He also did much to reinforce the military strength of the country. One aspect of his ambitious building programme was the fortification of key cities (1 Kings 9¹⁵). At Megiddo and elsewhere he maintained formidable forces of chariotry, an arm which at the time of the conquest had daunted the invading Israelites and of which David appears to have made no use, but which subsequently played an important part in Israelite military strength.

As Solomon's ambitious and lucrative trading activities did not prevent the impoverishment of his kingdom, so his political and military measures to strengthen it did not safeguard him against serious reverses. The capture of Damascus by Rezon (1 Kings 11²³⁻²⁵) must have been a hard blow at Israelite influence in Syria; and in Edom the activities of Hadad, a rebel prince, threatened Solomon's control of an important region. But the greatest source of weakness was internal. As subsequent events showed, Solomon's harsh and extortionate measures and his autocratic rule had provoked something more serious than discontent. His centralizing administrative policy was a serious blow to the old tribal or local administration and the traditional structure of Israelite society. But administrative unification could not produce true unity; nor could it stifle the will to revolt. Solomon's peculiar blend of astuteness and regal folly gave to Israel a golden age of prosperity and a legacy of division. His reputation for wisdom may well point to something more than his own quick-wittedness and command of proverbial lore. The range of Israel's international contacts in his time probably facilitated acquaintance with the cosmopolitan teaching of the ancient Near East and stimulated the cultural and intellectual life of the nation. But it is difficult to deny Solomon a high place in any representative list of wise fools who have occupied a throne.

Religious Developments

During the centuries which followed Joshua's invasion of Canaan, Israel's own way of life was confronted and influenced by Canaanite culture, and, in particular, by Canaanite religion, in three main ways. First, the Israelites had to adapt themselves to agricultural and urban life. Second, they had to live alongside Canaanite communities, for, as we have seen, considerable areas in the country continued to be inhabited or even controlled by Canaanites: the consolidation of the monarchy under David and Solomon involved the incorporation of such areas in the Israelite realm. Third, the establishment of the monarchy and the transformation of

Israel into a nation-state exposed Israel in new ways to the impact of Canaanite and other foreign influences. But it would be a mistake to suppose that Israel simply borrowed from or was corrupted by external religious beliefs and practices either at this or at any other period. In all the major religious developments which we can trace, we must reckon with Israel's own religion, which was not a dead tradition remaining passive while influenced from without, but an active, living faith, finding new modes of expression and unfolding its own truth in successive ages.

Our knowledge of the religion of Canaan is derived from three main sources. First, there is the Old Testament itself, in which many of the references are bitterly hostile, but in which there are also many indirect and incidental allusions. Second, there are references in ancient literature outside the Old Testament, but these (with the exception about to be noticed) are sketchy, sometimes obscure, and considerably later than the period to which they refer. Third, there are the results of archaeological discovery. Of outstanding importance are the excavations at Ras Shamra in northern Syria, the site of the ancient city of Ugarit, where there have been unearthed the remains of buildings (including temples), a wide range of objects of varied kinds, and also a very large number of texts, many of which are religious in character.[1] Though these date from the fifteenth or fourteenth century B.C. and come from a region considerably to the north of the land of Israel, they have so many points of contact with the Old Testament evidence about Israel's religious environment in Canaan that they may fairly be used to clarify and amplify the account of Canaanite religion which the Old Testament supplies. In them we have a considerable though fragmentary body of literary evidence.

The characteristics of Canaanite religion were those appropriate to a settled community, mainly agricultural but partly urban, in an area open to influences from various neighbouring regions. It was predominantly a nature religion. Its gods and goddesses were closely associated with the natural resources

[1] See *ANE*, pp. 92–132; *DOTT*, pp. 118–33.

and processes and with the cycle of the seasons; the theme of fertility played an important part in its myths and its cult; and the chief aim of its rituals was to ensure the continued well-being of the community and the productiveness of its land and its livestock.

Presentation of offering to El.

The chief god in the Canaanite pantheon was El. The word is a common term for 'god' in the Semitic languages, and may be used in a general sense of any divine being (cf. above p. 21); but it was also the personal name of the father and king of the gods, who presided over the divine assembly, and to whose authority other deities had to appeal. He was sometimes called 'the Bull-El'; and this was no doubt an indication of his connexion with animal vigour and fertility.

Nominally subordinate to El, but more active and in some ways more prominent, was Baal. The word 'baal' means 'owner', 'master', 'husband', and could be used as a common noun in quite general ways. It could also serve as the designation of any local deity. But it became the title and virtually the name of the young god of vegetation and fertility, who died, descended to the underworld, and rose again. He was also Hadad, the god of winter storm and rain. Representations of him have been discovered in which he is shown as a vigorous warrior, wearing a short skirt and a horned helmet, and carrying a club and a thunderbolt as a spear. He was involved in two major conflicts. The first was with Prince Sea, who is probably to be identified with the serpent Lotan (cf. Leviathan; Isa. 27[1]; Ps. 74[14]) and with the turbulent and destructive waters: in this, Baal was victorious. The second was with Mot, the god of death or of drought and barrenness. At first Baal had to yield to Mot and descend into the underworld; but later he was restored to life, fought again with his adversary, and won. These conflicts seem to have represented the alternation of the seasons in the agricultural year.

Closely associated with Baal was Anat, his sister and consort, who lamented him after his descent into the underworld, and whose vengeance on Mot is described in terms appropriate to the harvesting, winnowing, roasting, grinding, and sowing of corn (Mot probably represented, among other things, the ripe corn). She was a goddess of love and war. Other female deities were Asherah, the consort of El, and Astarte. Anat plays an important part in the Ras Shamra texts; but in the Old Testament her name occurs very infrequently and only incidentally (e.g. in the name of Jeremiah's birthplace, Anathoth). Asherah is mentioned in passages such as 1 Kings 18[19]; and the word is often applied in the Old Testament to a wooden cultic object. The name Astarte appears often in the Old Testament in the distorted form 'Ashtoreth' (plural, 'Ashtaroth'),[1] and seems to be used as a generic term for female fertility

[1] In 'Ashtoreth' the vowels of 'bosheth' (shame) have been inserted for religious reasons: cf. above, p. 53, n. 1.

Baal of the Lightning.

deities (see, e.g., 1 Sam. 7$^{3f.}$). It is impossible to distinguish sharply between these goddesses. All three are forms of the great goddess of love, motherhood, and war. Clearly their place in the fertility cult was an important one.

That cult was no mere formalism. It was concerned with the realities on which the life of the community depended. The due representation in word (the recital of the myth) and act (the dramatic symbolism of the ritual) was believed to be a potent means of maintaining the ordered harmony of nature and of the life of the community.

A religion of this kind presented a sharp challenge to the faith which the Israelite invaders brought with them. The Mosaic religion had as its setting the life of the semi-nomad, not that of the farmer. More important, its historical character was in marked contrast to the nature religion of Canaan. This contrast was to prove decisive. But there was a long and some-times painful process of adjustment and assimilation; and it is important not to oversimplify our picture either of the process or of the issues which were at stake.

Some conservative groups seem to have held that loyalty to Yahweh was incompatible with the Canaanite way of life. They maintained the old desert ways, refusing to live in houses, and to grow cereals and cultivate the vine. Such an attitude was probably easier to maintain east of Jordan, or in the far south, than in other parts of the country; but we find it in Judah as late as the sixth century B.C. (Jer. 35). For most, however, some kind of adjustment had to be made. Many, no doubt, adopted the worship of the Canaanite deities while continuing to recognize Yahweh as pre-eminently *the* God of the tribal con-federacy, to whom they could turn in times of special need or crisis. Or, again, He might be worshipped as if He were a Canaanite deity, at sanctuaries used for generations by the Canaanites and by means of rituals resembling those of Canaan, at festivals which were closely linked with agricultural life. Both the terms 'El' and 'Baal' were applied to Yahweh. As we have seen (above, pp. 20 f.), the former appears in various compound titles in the stories about the patriarchs; and there is no indication that the application of it to Yahweh aroused opposition or criticism. 'Baal', too, was used as a title of Yahweh. Saul, Jonathan, and David all had children in whose names 'Baal' was a component (1 Chron. 8[33 f.], 14[7]); and it may be taken as certain that the reference was to the God of Israel. But at a later period hostility to the fertility cult had become so intense that the very word 'Baal' was regarded as a mark of apostasy. The prophet Hosea says not only that Israel must learn that the produce of the land comes from Yahweh and not from the local Baals (Hos. 2[5, 8]), but also that she must call Yahweh 'Ishi' (my Husband) and not 'Baali' (my Baal) (2[16 f.]). It is significant that in some passages the component 'Baal' in the names just referred to has been suppressed (2 Sam. 2[8], 4[4], 5[16]; cf. above, p. 53, n. 1).

We must beware of thinking too abstractly about the assimilation or rejection of Canaanite ways. In the process an important part was necessarily played by the *places*, *objects*, *acts*, and *times* which were associated with worship, and by the *persons* who were held to have a special relationship to the deity.

We have already seen that Solomon's Temple was constructed on a plan current in Phoenicia and Canaan. It may also have been on an ancient holy site. In other parts of the land, ancient sanctuaries had been taken over by the Israelites. These would vary considerably in size and complexity, from the simple holy place associated with a stream or spring or a sacred tree, or located on a sacred height, to more elaborate and extensive sanctuaries. The special use of the term 'high place' (e.g. 1 Kings 3[4]) for a cultic site appears to imply a location on a hill-top; but this is not necessarily suggested by the Hebrew word *bamah* (plur. *bamoth*). A 'high place' could be in a city (e.g. 2 Kings 17[9, 11]) or even in a valley (Jer. 7[31]). Presumably it was some sort of artificial or natural mound or raised platform. It has been suggested that originally the 'high place' was a burial mound and that it was connected with the cult of ancestors. The importance of sanctuaries such as Gilgal, Bethel, Gibeon, Ophrah, Mizpah, and Dan, is reflected by the place either of note or of notoriety which they occupy in the Old Testament traditions. But, above all, the sanctuary which housed the Ark was the religious rallying place of the tribal confederacy. In the period immediately before the establishment of the monarchy, this was Shiloh, as the opening chapters of 1 Samuel show. It represented something of more than

Standing Stones at Gezer.

local significance: the covenant bond which linked the tribes to each other and to the God who had brought their ancestors out of Egyptian bondage.

The three most important items of Canaanite sanctuary equipment were the altar, the stone pillar (*maṣṣebhah*) which represented the male divinity and the wooden pole (*'asherah*) which represented the female divinity (though Jer. 2²⁷ appears to suggest the reverse). The second and third of these came under the condemnation of Israelite reformers, though in some early passages the pillar was regarded as an appropriate token of the divine presence (e.g. Gen. 28¹⁸).

A consideration of the acts performed at the sanctuaries cannot easily be separated from some account of the sacred seasons or festivals; but we may note the three main types of sacrifice and offering which belong to this period. The peace-offerings (sometimes simply called 'sacrifices') established or renewed communion between the deity and the worshippers, who partook of a sacrificial meal after the offering had been made. The burnt offering was given wholly to the deity; no part of it was consumed by the worshippers. Of cereal offerings perhaps the most notable instance at this period was the first-fruits. On the whole, the references to sacrificial practice in passages which may confidently be dated early are not very explicit or detailed. Precise directions in such matters are given in the later sources; but it should not be assumed that late sources never record ancient practice. One feature of Canaanite sacrificial practice which roused strong opposition in strict Yahwistic circles was human sacrifice. This, however, was not unknown in Israel (cf. Judges 11[34-40]) and appears to have been particularly common in the seventh century.

According to the general view, three great festivals belonged to the structure of agricultural life in Canaan: the feast of Unleavened Bread at the beginning of the barley harvest, the feast of Weeks at the wheat harvest, and the feast of Ingathering (in later passages called the feast of Tabernacles or Booths). The fact that they occurred at significant seasons in the farmer's year is an indication of the vital connexion between religious observance and life and work.

In the period between the settlement and the Exile, the third of these three festivals seems to have been the most important. It fell at the close of the agricultural year (September/October), when the fruits were gathered in. It was an occasion for rejoicing and merrymaking; and the tabernacles or booths from which the festival takes one of its names were the rough shelters which were used by the worshippers during their celebrations out in the fields. There is here a pointer in two directions: first, to the agricultural associations of the festival, and second, to the connexion which it came to have with the historic deliverance of Israel from Egypt. For (to take the second point first) the reason which is given in Lev. 23[43] for the command to 'dwell in booths for seven days' is 'that

Gezer calendar (probably tenth or ninth century B.C.). Listing phases of the agricultural year.

your generations may know that I made the people of Israel dwell in booths, when I brought them out of the land of Egypt'. The 'booths' were, in fact, very different from the tents of desert wanderers: they were constructed from the branches of trees. But the association of them with the wilderness experience of the Israelites is an instance of a tendency which appears frequently in Old Testament religion, to relate what was taken from other contexts to the historical events which were the foundation of Israel's religion. It has commonly been maintained that the association of the feast of Unleavened Bread with Passover and with the Exodus is another example of this tendency. Similarly, at a later period the feast of Weeks, which in the Old Testament laws is unmistakably a harvest festival, came to be regarded as a celebration of the institution of the Covenant at Sinai and the giving of the Law. On the other hand, the agricultural context of the feast of Tabernacles is plain; and it has been suggested that the 'booths' were originally not only the shelters used by the worshippers, but representations of the bridal bower in which was celebrated the sacred marriage (*hieros gamos*) between the fertility god and his consort. This is far removed from the spirit of the religion of Israel, whose God, Yahweh, was not a fertility deity and had no consort. Yet the festival was taken over by Israel; and even if it was modified and reinterpreted, the question remains what its original character was and how much of its structure and content found a place in Israelite practice. This is one of the most disputed questions in the study of Israel's religion. It has been argued that in the pre-exilic period this was a New Year Festival, the aim of which was the annual renewal of the natural and social order. Parallels have been adduced from other religions of the ancient Near East, particularly from Baby-lonian and Ugaritic texts, relating to the conflicts waged by deities such as Baal with the powers of chaos and death, their ultimate triumph, and the blessing which ensued. So it has been maintained, Israel celebrated annually Yahweh's triumph over the powers of chaos, and acclaimed Him as King when, on the great day of the autumnal festival, He took His place on the throne, and renewed His covenant with His people. Yahweh's lordship over nature meant the recapitulation of the work of creation, the restraining of destructive forces, the promoting of fertility, and, in particular, the coming of the autumnal rains which made it possible for the farmer to start the work of a new year. His lordship over men implied His coming as Judge and Saviour of His people, to repel the attacks of their enemies and to promote internal harmony. It has been further argued that this pattern of conflict, triumph, and renewal, experienced annually in the festal worship, provided the structure and content of the hope of a new age of restoration which was cherished at a later period in Israel's religious history. Criticism of this theory has fastened particularly on the lack of direct evidence in the Old Testament laws, and on the fact that the theory is based partly on non-Israelite usage, and partly on inference from Old Testament passages such as the so-called Enthronement Psalms (e.g. 47, 93, 95–100), the interpretation of which is open to dispute. Nevertheless, the cumulative evidence for the main outlines of the theory is strong. Several passages in the Old Testament indicate that the mythical theme of the conflict with the destructive waters of chaos was applied to Yahweh, and fused with Israel's own historical tradition of Yahweh's act in dividing the waters of the Red Sea (e.g. Ps. 74^{12-14}, 89^{10}; Isa. $51^{9\,f.}$: no clear distinction is made between the sea or the chaos waters on the one hand, and on the other, the chaos monster, referred to variously as 'the dragon', 'Leviathan', and 'Rahab'). The historical character of Israel's religion transformed what was taken over from Canaanite festal worship and its accompanying mythology; but at the same time the conception of Yahweh's lordship over nature was enriched. One of the supremely important elements in Israelite tradition was the idea of the covenant between Yahweh and His people; and it has been argued that the autumnal festival was primarily concerned with the renewal of the covenant. That the covenant played a part in its observance we need not doubt (though later the Feast of Weeks was the festival which commemorated the covenant); and accordingly, one effect of the annual celebration of the festival would be to strengthen the bond which held the tribal confederacy together.

As we have seen, on one widely held view, the theme of the Kingship of Yahweh was central in this festival, whether in an enactment of His enthronement, or in an acknowledgement and celebration of His perennial kingly rule. It is also claimed that the festival had a special significance for the reigning king and the royal house, since the king was thought of as in a special sense the representative of God to the people and of the people to God. Like the character of the autumnal festival, the religious status of the king has been a hotly disputed question in recent study of Old Testament religion. Elsewhere in the ancient Near East, kings exercised important religious functions and were believed to have a specially close relationship with the divine world, and to be the channel by which the gods bestowed their blessings on the community. But there was no uniformity of belief and practice. In Egypt, indeed, an unmistakably divine character was ascribed to the Pharaoh, who was held to be the incarnate son of the deity. Elsewhere (e.g. in Mesopotamia), the special position of the king was expressed in terms of the choice and adoption of him by the deity. In general, however, his status as representative and intermediary is evident. Since he embodied in his own person the life of the community of which he was head (cf. what was said above about the patriarchs, p. 15), his vigour and well-being were of vital importance for that community.

In adopting the monarchy Israel took over from her Canaanite environment an institution which was not only political but religious; and it is not surprising that kings exercised an important influence on the religious life of the nation and at times performed cultic functions. The fact that kings such as David, Solomon, Jeroboam I (of Northern Israel), Joash, Ahaz, Hezekiah, and Josiah planned or built sanctuaries, or gave them special patronage, or took active measures to alter their furnishings or reform their worship, or appointed their priests is perhaps only what might be expected of the heads of a state in which religion was an official national activity. But Israelite kings themselves performed important cultic acts. Sacrifices were offered by Saul at Gilgal (1 Sam. $13^{9f.}$) and by Solomon at Gibeon and at the dedication of the Temple at Jerusalem (1 Kings 3^4, $8^{5,\ 63f.}$). David both sacrificed (2 Sam. $6^{13,\ 17f.}$) and performed a cultic dance before the Ark, wearing a priestly vestment (2 Sam. 6[14]). Further, the king was 'the LORD's anointed', 'the anointed of Yahweh' (e.g. 1 Sam. 24[6], 26[11]), and, as such, sacrosanct. Anointing (1 Sam. 10[1], 16[13]; 1 Kings 1[45]) was a sacramental act which conferred a special status and conveyed divine power. How important for the well-being of the community was the presence of a person so endowed is vividly expressed in the title 'the lamp of Israel' (2 Sam. 21[17]; cf. Lam. 4[20]), or at greater length in a psalm such as 72. The intimate relationship between the king and Yahweh is illustrated by Ps. 2, which is in all probability a psalm for a king's accession. The words (v. 7) 'You are my son, today I have begotten you' (or, rather, '*I* have begotten you today') appear to indicate the belief that the king became at his accession the adopted son of Yahweh.

But here, as elsewhere, Israel brought to what she borrowed the transforming influence of her own inheritance. Kingship was not merely an alien intrusion. As the period of the Judges prepared the way for the establishment of monarchy, so the monarchy in its historical origins, and always at least ideally, continued the tradition of the charismatic leadership exercised by the Judges, the men raised up and endowed by Yahweh for the deliverance of His people. Further, the relationship of the Davidic[1] royal house to Yahweh was interpreted in terms of the covenant idea. The idea, though not the actual word 'covenant', is present in 2 Sam. 7, where, in response to David's desire to build a house for the Ark (i.e. a temple), Yahweh replies that He will build a house (i.e. a dynasty) for David. The same thought appears with particular emphasis and clarity in Ps. 89 (see especially vv. 3 f., 19–37); and there the actual word 'covenant' is used. Yahweh's gracious choice, His word of promise and enduring faithfulness, and on man's side the obligation of loyalty and obedience: in all of these

[1] The qualification 'Davidic' is important. It is a somewhat artificial oversimplification of the facts to suppose that there was only one conception of monarchy in Israel. Not only did the passage of time bring development; but the Northern Kingdom of Israel, which came into existence shortly after Solomon's death, had no stable dynasty and no royal tradition comparable to that of the Judaean house of David.

there is a parallel between the relationship of Yahweh to His people and that of Yahweh to the house of David. Or, to use a somewhat different geometrical figure, Yahweh's covenant with the house of David is a smaller concentric circle within the larger circle of His covenant with Israel. The presence of a scion of David was an outward token of the continuing goodness of Yahweh to His people.

At a later period, when the hope of restoration developed, one element in it was the expectation of such a scion of David, an 'anointed (one) of Yahweh'. It is from the Hebrew word for 'anointed one', *mashiᵃḥ*, that our word 'Messiah' is ultimately derived. If it is true, as has been maintained, that the autumnal festival had a particular importance for the royal house, and if, further, the festival provided the structure and content of the later hope of restoration (see above, p. 73), then it is not surprising that the expectation of a Messiah was an element in the pattern of that hope. But it was only one element.

As a sacral personage, the king was associated with special occasions and activities. The regular and routine observance of the cult were the responsibility of the priests; but what were the precise functions of the priesthood, and who were entitled to exercise these functions, are questions to which different answers must be given for different periods of Old Testament history. For the time between the settlement and the end of Solomon's reign, the evidence is unfortunately scanty.

In Deut. 33[8-10] three main functions are ascribed to Levi, who here represents the Israelite priesthood. (1) 'Give to Levi thy Thummim, and thy Urim to thy godly one' means that the priest ascertains the divine will by the manipulation of the sacred dice, the Urim and Thummim. (2) 'They shall teach Jacob thy ordinances, and Israel thy law' refers to the teaching office of the priesthood. (3) 'They shall put incense before thee, and whole burnt offering upon thy altar' describes the priestly service of God at the altar.

It is probably significant that the giving of oracles and instruction in the law precede altar service. A different order

appears in the summary of priestly functions in 1 Sam. 2²⁸: 'to
go up to my altar, to burn incense, to wear an ephod before
me'; but there, too, there is probably a reference to the dis-
cerning of the divine will. The ephod is described sometimes as
a priestly vestment (1 Sam. 2¹⁸; 2 Sam. 6¹⁴) and sometimes as
a cultic object (Judges 8²⁷). It also appears to have had some
association with the giving of oracles (1 Sam. 23⁹⁻¹²). We cannot
be sure of its precise character, or, indeed, whether there was
only one kind of ephod; but since in the account of the high
priest's vestments (Exod. 28) we are told that there was
attached a 'breastpiece' which was in effect a pouch to hold the
Urim and Thummim (the oracular dice), the ephod which
was a vestment was not necessarily distinct from the ephod
which was associated with the giving of oracles.

The teaching office of the priesthood was probably of far
greater importance than is commonly realized; and it was
certainly wider in scope than is suggested by the English words
'ordinances' and 'law' in Deut. 33¹⁰. It involved not only
giving decisions on moot points and guidance about ritual
usage, but general instruction about the will of Yahweh.
Although the Hebrew word *torah* (law) could be used of the
revelation imparted through the prophet (Isa. 1¹⁰), it was used
with peculiar appropriateness of priestly instruction. In Jer.
18¹⁸ the *torah* of the priest is bracketed with the counsel of the
wise man and the word of the prophet. Priests are denounced
because they make their teaching office a source of profit
(Mic. 3¹¹; cf. Mal. 2⁷ᶠ·).

The burning of incense and the offering of sacrifice are
linked together because of their association with the altar.
In sacrifice, the specific task of the priest was not the actual
slaughtering of the victim (though he might, in some circumstances, do this), but the sprinkling of the blood and the laying
on the altar of what was given to God. The function came to be
of special importance in later times, when the giving of oracles
by priests had ceased and their teaching office was less prominent. In the early period, though sacrifice was one of the
functions of the priests, it was not restricted to them. Sacrifice

was offered by Gideon (Judges $6^{25\text{ ff.}}$) and by Samson's father, Manoah (Judges 13^{16-23}).

The Levites had a special relationship to the priestly office and the service of the sanctuary. In Deuteronomy, though the position is not entirely clear, priests and Levites seem to be equated (e.g. Deut. 18^{1-5}). In the Priestly Source, the specifically priestly functions are reserved for those Levites who are descended from Aaron, whereas the remaining Levites are to be responsible for other forms of service in the sanctuary (Num. 18^{1-7}). An even narrower limitation of the service of the altar is prescribed in the book of Ezekiel, where full priestly status is reserved for the Zadokites (Ezek. 40^{46}). But in the period with which we are at present concerned, not all priests were Levites. Though he was an Ephraimite (1 Sam. 1^1), Samuel was a priest at the sanctuary at Shiloh; and after it was destroyed he continued to perform priestly functions in different parts of the country. David, who was of the tribe of Judah, had sons who were priests (2 Sam. 8^{18}). Judges 17 tells of an Ephraimite named Micah who established a private sanctuary and installed one of his sons as priest (v. 5); but when a Levite turned up, Micah immediately took the opportunity of employing him (vv. $7-13$). It appears then, that at least until the early period of the monarchy, priesthood was not confined to the Levites, though Levitical priests were preferred.

The biblical traditions about the origins and early history of the Levites are difficult to interpret. On the one hand, they are represented as having lost their tribal cohesion and become dispersed in Israel (Gen. 49^{5-7}) because of an atrocity committed in pre-Mosaic times. On the other hand, the priestly office is said to have been assigned to them because of their attitude to idolatry and apostasy (Exod. 32^{26-29}; Deut. 33^{8-10}). The suggestion has been made that there was a non-priestly tribe of Levi, which was distinct from the Levites who specialized in cultic functions, but which in effect ceased to exist as a recognizable entity. Where the evidence is fragmentary and obscure, it is unwise to be dogmatic; but it seems to be neither impossible nor improbable that there was in fact one tribe of Levi, which, at an early period, came to specialize in cultic functions, and later assimilated other elements who shared in these functions at various sanctuaries.

We have already seen that the establishment of the monarchy was in itself an important religious development, and that the choice of the royal city, and the building of the royal sanctuary, had far-reaching effects on national life and religious practice. These developments also had both immediate and long-term consequences for the priesthood. After the Philistines had destroyed Shiloh and its temple, an important community of priests was presided over at Nob by Ahimelech, a descendant of Eli; but it was almost completely wiped out by Saul, because Ahimelech had helped David (1 Sam. 21, 22). The sole survivor, Abiathar, joined David in outlawry and served as his priest. After the capture of Jerusalem, another priest, Zadok, appears in David's entourage. The priests who were attached to the court and the royal sanctuary would naturally have a special prestige and authority. At the beginning of Solomon's reign, Abiathar fell from favour and was dismissed. Thereafter, the Zadokite priests, like the Temple at which they served, were to exercise for centuries a role of increasing importance.

The period of the Philistine wars and of the rise of the monarchy also provides us with the first clear evidence of the activity of the prophets. In 1 Sam. 10$^{5-13}$ there is a brief description of a group of prophets whom Saul met shortly after he had been anointed by Samuel. In the previous chapter, Samuel himself is called a 'man of God' (9$^{6-8, 10}$) and a 'seer' (9$^{11, 18}$); and in an explanatory parenthesis we are told, 'he who is now called a prophet was formerly called a seer' (9^9). The meaning of this seemingly simple statement has been the subject of much dispute. Some scholars have thought that there must at one time have been two quite distinct types of holy men in Israel, described by these two terms: the prophets, who were gregarious and were subject to fits of religious frenzy; and the seers, who were solitaries, and received and imparted their revelations in a calm state of mind. But the available evidence makes it impossible to draw such sharp distinctions;

and the parenthesis itself says simply that the term 'seer' had come to be replaced by 'prophet'. Basically, the term 'seer' indicates the visual experiences of the man of God.[1] He had the ability to see what others could not see. But the word does not necessarily have this limitation: the verb 'to see' could be used of a prophet's auditory experience. 'Prophet' represents the Hebrew word *nabhi'*. It used to be thought that this word was connected with a verb meaning 'to bubble' or 'pour forth' and that it referred to the uncontrolled speech of the prophet when in a frenzy. It is much more likely that it is connected with an Akkadian verb *nabu*, meaning 'to call', and that *nabhi'* means either 'one who is called', or 'one who calls out'.

But terminology and etymology are probably less useful as guides to the nature of prophecy than the records of what the prophets were and did. The description in 1 Sam. 10[5-13] indicates three features of the prophetic movement: its gregarious character; the so-called ecstatic element in the prophetic experience; and its relationship to the cult and the sanctuaries.

Saul met a group of prophets; and it seems likely that such companies were a common feature of religious life in Israel at the time. Samuel himself seems to have had close connexions with them (1 Sam. 19[20]); and in later periods we learn of prophetic communities settled in different parts of the country. The expression 'sons of the prophets' refers to such communities (1 Kings 20[35]; 2 Kings 2[3, 5, 7, 15], 4[1, 38]). They were not celibate; but shared a common life for the exercise of their prophetic functions.

The statement that the prophets were 'prophesying' (1 Sam. 10[5]) does not mean that they were uttering some religious message or prediction. It denotes the outward expression in bodily movement and incoherent utterance which resulted from an abnormal psychological condition. This state, which is commonly called ecstasy, could be induced by artificial means, such as music (2 Kings 3[15]). The stories of Saul's

[1] The Hebrew words *ro'eh* and *ḥozeh*, both of which mean 'seer', seem to be virtually or wholly identical in meaning, although attempts have been made to distinguish between them.

experiences indicate that the frenzy could have an infectious character (1 Sam. 10⁶, ¹⁰, 19¹⁸⁻²⁴). But Yahweh was held to be the source of the prophets' inspiration. This is indicated by expressions such as 'the hand of the LORD was upon . . .', or, 'the spirit of the LORD came upon . . .' (1 Sam. 10⁶, 19²⁰; 1 Kings 18⁴⁶; 2 Kings 3¹⁵). In this experience of possession by the spirit, the prophets resemble the Judges, who were endowed with supernormal gifts of courage, military skill, and leadership.

The fact that the prophets whom Saul met were coming down from the high place is probably significant. Prophets appear to have had important connexions with the cultic life of ancient Israel. This has often been overlooked or denied, partly because of the common but misleading antithesis of prophet and priest, and partly because of the denunciations of sacrificial practices and other cultic observances in some of the prophetic books. But it seems highly probable that some prophets, at least, had an important part to play in cultic activity and in connexion with the sanctuaries. Part of the evidence for this is derived from other religions of the ancient Near East, where similar types of holy men (diviners and the like) are found associated with sanctuaries. But there is also evidence in the Old Testament itself. Samuel was both prophet and priest (see, especially, 1 Sam. 3¹⁹⁻²¹). It appears that prophets were to some extent under the authority of the Temple priesthood at Jerusalem in the time of Jeremiah (Jer. 29²⁶). It has also been maintained that some passages in the Psalms are oracles which were uttered by the sanctuary prophets, and that in the post-exilic period the successors of these prophets became Levitical choirs in the Temple service.[1]

There is one important and illuminating connexion between prophetic and cultic practice. In both, we find the conception of the creative or destructive power inherent in certain words and actions. We have already seen (above, p. 67) that in the cult both what was said and what was done had creative

[1] Notice the use of the word 'prophesy' in 1 Chron. 25¹⁻³, and the prophetic word uttered by a Levite in 2 Chron. 20¹⁴ᶠᶠ.

power; and similarly it is clear that what the prophets said was a communication not only of information but of power, and, further, that they sometimes performed actions which were intended to effect what they represented. In 1 Kings 22[12], for instance, when the prophets say, 'Go up to Ramoth-gilead and triumph', the intention is that their words, charged with supernatural potency, will help to bring about victory; and this is reinforced by the imitative action of pushing with horns (v. 11, cf. 2 Kings 13[14-19]; Isa. 20). Such actions are sometimes called dramatic symbolism, or acted prophecy. The world of thought to which they belong may seem to be dangerously akin to magic, in which man manipulates the resources of supernatural power for his own ends. But, though this danger was no doubt present, there is clearly a different strain in Hebrew prophecy. This is clearly seen in the incident just referred to (1 Kings 22), when the prophet Micaiah says: 'What the LORD says to me, that I will speak' (v. 14). Here the will of God is supreme; and there is no question of adapting prophetic speech or action to national policy or royal whim. What the prophet says and does is, indeed, charged with power; but it is the power of Yahweh's own will. This concern with the active will of Yahweh is central in the classical tradition of Hebrew prophecy.

It would be outrunning the evidence to suppose that all prophets in ancient Israel were members of prophetic communities, experienced ecstatic frenzy, and had definite cultic functions; but these may be taken as general characteristics of the prophetic movement. Parallels to these features can be found in other religions of the ancient Near East. The Old Testament itself refers to the prophets of deities other than Yahweh, such as the prophets of Baal mentioned in the Elijah story (1 Kings 18). An Egyptian document which describes the travels and adventures of a certain Wenamon in the eleventh century, tells how, when the ruler of Byblos was offering sacrifice, a young man who was present went into an ecstatic frenzy and uttered an inspired oracle. In Mesopotamian religion there was a type of priest-prophet known as *baru*; and we also hear in the Mari documents of the *maḫḫu*, an envoy of the gods, who communicated the divine will in terminology reminiscent of that found in the Hebrew prophetic literature. The existence of these similarities has, not unnaturally, prompted theories about the possible origin of Hebrew prophecy, which has sometimes been sought as far afield as Asia Minor or Thrace. Clearly, if some non-Israelite origin is to be assumed, there is no need to go outside Israel's Canaanite environment, where similar phenomena were to be found; and we have already seen that in other aspects of culture and religion Canaan exercised an influence on Israel. But, whereas agricultural festivals may naturally be supposed to have been taken over from an agrarian culture, the phenomena of prophecy are so widespread, both historically and geographically, that the theory that they are part of Canaan's legacy to Israel is not so cogent.

In any event, prophecy in Israel became something markedly different from anything known to us in Israel's environment. This will become evident when we consider the so-called classical prophets of the eighth and later centuries. But it is wrong to make a sharp distinction between them and the earlier prophecy. Whatever influences were brought to bear on Israel's religion from outside, faith in the Saviour God, Yahweh, who brought Israel out of Egypt, remained central. Prophecy, for all its seeming extravagances and bizarre manifestations, was concerned with the will of Yahweh, not only as something that men were required to do, but as a force with which they had to reckon in the history of their own day. Elsewhere, the holy man might try to make events conform to human desires, and the diviner might try to satisfy human curiosity about the future; but in Israel there were prophets, who, as messengers of Yahweh, were heralds and interpreters of what He was doing and what He was about to do.

In the period which we have surveyed, there was rapid and drastic development of Israel's life and culture, in which her religious inheritance was both tested and enriched. The establishment of the monarchy, particularly under David and

Solomon, made possible the consolidation of many of these varied influences. But the period of prosperity, security, and unity was brief. Centuries of crisis were ahead; and during that time, the prophetic movement, which first clearly emerged in the crisis of Philistine oppression, was to exercise a decisive role.

V

FROM THE DIVISION OF
THE KINGDOM TO THE EXILE

During the period of rather less than three and a half centuries which separated the end of Solomon's reign (922 B.C.) from the fall of Jerusalem (587 B.C.), the kingdom which David had established passed from the position of security, power, and affluence which he had won for it, through division, decline, partial and temporary rehabilitation, and further decline, to extinction as an independent political unit. The story is a complex one. It involves the relations to each other of the two parts into which the kingdom split, and also the impact upon them of nearer neighbours, such as the Syrian or Aramaean kingdom to the north, or more distant and powerful countries such as Assyria and Egypt, and, at the end of the period, the neo-Babylonian empire. It also includes the significant economic and social changes by which the structure of Israelite communal life was affected, and, in the religious field, the interplay of alien religious influences and internal movements of reform. Decisive events mark the beginning and the end of the period; but between these two extremes, clear lines of demarcation are considerably less easy to draw. The following main phases will be considered in our survey: the division of the kingdom and its immediate aftermath (922–876 B.C.); from the rise of Omri to the accession of Jeroboam II (876–786 B.C.); from the reigns of Jeroboam II of Israel and Uzziah of Judah to the death of Hezekiah of Judah (786–687 B.C.); from the accession of Manasseh to the fall of Jerusalem (687–587 B.C.).[1]

[1] The dates given below for the reigns of the kings of Israel and Judah do not always correspond with the durations of their reigns as stated in the Books of Kings. The complex chronological problems involved cannot be

The Division of the Kingdom and its Immediate Aftermath
(922–876 B.C.)

The king who built the Temple undermined the kingdom. Solomon's policies had in them, as we have seen, an element of tyranny which had already provoked discontent before his reign ended. When his son Rehoboam (922–915 B.C.) came to the throne, there was not, so far as we know, a struggle for power such as preceded Solomon's accession; but there was an immediate plea for less oppressive policies (1 Kings 12[1-16]). This plea was presented at an assembly at Shechem, to which Rehoboam had come to be acknowledged as king. The place and its associations are important. Clearly the traditions and self-consciousness of the northern tribes were far from dormant. The tendencies to disunity which were present even in David's reign were not merely latent; and as David, king of Judah, had become king over all Israel by covenant, so now allegiance was offered on certain conditions. The conditions were refused and the allegiance was withdrawn. Apparently a further important factor in this crisis was incitement by a representative of the prophetic movement. The prophet Ahijah had earlier prompted the Ephraimite, Jeroboam, a senior official in charge of the *corvée*, to rebel against Solomon (1 Kings 11[28 ff.]);[1] and when the northern tribes revolted against Rehoboam, it was Jeroboam (922–901 B.C.) whom they chose as their king (1 Kings 12[20]). This prophetic intervention is a striking instance of the active involvement of the prophetic movement in political affairs, and probably also an expression of a critical attitude to the kind of monarchy which had been established.

Thus, long-standing tension between north and south, the harsh measures of Solomon, prophetic influence, and the obduracy of Rehoboam combined to divide the kingdom. David's dominions and spheres of influence beyond the bounds

---

discussed here. In general, though not in every particular, the system followed is that adopted in J. Bright, *A History of Israel* (London, 1960).

[1] Notice the interesting piece of dramatic symbolism in the tearing of a garment into twelve pieces (cf. above, p. 82).

of Israel and Judah, already somewhat diminished, were now lost. There remained to David's successors only the tiny southern state of Judah. The other heir of the Israelite inheritance was the northern kingdom of Israel, sometimes called Ephraim, the name of the most powerful of the northern tribes. Between these two states there were important differences which are not immediately evident from the biblical record. In area, fertility, population, and military potential, the northern kingdom was far superior to Judah. When the two kingdoms were in alliance with each other, Judah, as the weaker and poorer, was inevitably the junior partner, if not, indeed, the vassal. The advantages which the northern kingdom enjoyed, together with the fact that its territories were crossed by important lines of communication, made it both a more desirable ally and a more tempting prey than Judah. Judah's political insignificance and her isolated position made it easier for her rulers, if they so wished, to avoid entanglement in international politics; but for the northern kingdom this was out of the question. This doubtless helps to account for the fact that Judah survived for some 135 years after the fall of the northern kingdom.

There is also a striking difference between the history of the two monarchies. Until the fall of Jerusalem, the house of David continued to reign in Jerusalem, unless the short period of Athaliah's rule is regarded as an exception (2 Kings 11$^{1-16}$). But in the north there were several changes of dynasty, which were brought about by *coups d'état* and assassinations. Judah's internal history was, therefore, more stable.

Judah also possessed the new capital, Jerusalem, with its magnificent Temple, in which the Ark was enshrined. The Temple, the city, and the Davidic dynasty came later to have special significance in the theology of election and covenant; but even at this stage there was a special advantage to the royal house of Judah in having possession of the Temple. The practice of worship there would reinforce loyalty to the house of David. This fact provided the motive for the religious measures which were adopted by Jeroboam. The account of these which is given in 1 Kings 12²⁶⁻³³ is coloured by the Jerusalem tradition and by the standards embodied in Deuteronomy. But it is understandable that Jeroboam should have given royal patronage to the ancient sanctuaries at Bethel and Dan. The bull images which he set up there are now commonly held to have been intended as pedestals of the invisible Yahweh, and not as idols, though it was no doubt perilously easy for them to be popularly regarded as idols. Why he should have instituted an autumnal festival in the eighth month rather than the seventh is not clear; but his interest in it is understandable if its southern counterpart had a special significance for the royal house (see above, p. 74). The criticism in 1 Kings 12²⁶⁻³³ of Jeroboam's religious policy is in line with the attitude adopted throughout the Deuteronomistic history. All kings of either kingdom who tolerated or encouraged Canaanite or other alien religious practices are condemned; and since the Deuteronomic ideal included not only the purification of the cult but also its centralization at the one legitimate sanctuary, Jerusalem, failure to suppress the local shrines was regarded as a defect in the religious policy of kings who were otherwise commended. In this the kings of the northern kingdom were inevitably found wanting.

At first, Shechem served as Jeroboam's administrative centre. East of Jordan, he built Penuel, possibly as a second capital city (1 Kings 12²⁵); but finally he made Tirzah his capital. No major campaign seems to have been undertaken between the two Israelite states; but border fighting resulted in Judah's retaining the territory of Benjamin, which served as a useful bulwark to the north of Jerusalem. Both countries had to face aggression from another quarter. During Solomon's reign, Egypt had harboured both Jeroboam and the Edomite prince, Hadad (1 Kings 11¹⁴⁻²², ⁴⁰). Shishak, the first Pharaoh of the Twenty-second Dynasty, aimed at re-establishing Egyptian influence in western Asia. He invaded Palestine and compelled Rehoboam to pay immense tribute (1 Kings 14²⁵⁻²⁸). His own account of the campaign, recorded in a hieroglyphic inscription at Karnak, shows that he also did serious damage to the northern kingdom, probably by an encircling march eastwards, through part of Transjordania, back across the plain of Esdraelon, and finally, southwards by the coastal plain.

Hostility between the two kingdoms continued during the reigns of Rehoboam's two immediate successors, Abijam

List of Palestinian and Syrian towns captured by Sheshonk I (Shishak).

(915–913) and Asa (913–873). In the northern kingdom, internal dissension was soon evident. After a short reign of only two years (901–900), Jeroboam's successor, Nadab, was assassinated by Baasha while he was besieging Gibbethon. Baasha exterminated the entire family of Jeroboam and established himself as king (900–877). Similarly, when Baasha himself died, his son, Elah, was assassinated by Zimri, a senior army officer. Zimri's own reign lasted only a week. The Israelite force which was besieging Gibbethon acclaimed another officer, Omri, as king and marched against the capital, Tirzah, where Zimri, abandoning hope of resistance, set fire to his own palace and died in the conflagration. This, however, did not put an end to civil strife. For a time Omri had to reckon with a rival, Tibni, who received considerable popular support. But Omri's final vindication of his claim to the throne of the northern kingdom ended the period of insecurity and inaugurated one of strong rule, renewed prosperity, influence abroad, and of critical religious conflict at home. It was also a time of improved relations between the two Hebrew kingdoms.

From the Rise of Omri to the Accession of Jeroboam II (876–786 B.C.)

Omri's reign (876–869) is almost summarily dismissed in the biblical record (1 Kings 16²³⁻²⁸); but archaeological evidence, extra-biblical texts, and legitimate inference from the narratives in Kings indicate that his political achievement was considerable. Even after his dynasty had been destroyed, Assyrian records referred to Israel as 'the House of Omri'; and the famous Moabite stone, which records the success of Mesha, king of Moab, in freeing his country from Israelite domination, makes it clear that Omri had subjugated Moab. The part which his son, Ahab, played in the international scene was in some measure the result of Omri's prowess. It was probably Omri who was responsible for the marriage of Ahab with Jezebel, daughter of the Phoenician king, Ethbaal or Ittobaal (1 Kings 16³¹). He founded the new capital, Samaria (replacing Tirzah), in a magnificent situation (1 Kings 16²⁴); and excavations have revealed the strength of the defences and the beauty of the ivories with which the palace was adorned (1 Kings 22³⁹). In his work of stabilization after a period of weakness and internal strife, his choice of a new capital, and his policy of alliance and co-operation with Phoenicia, Omri may with some justice be regarded as the David of the northern kingdom. But the altered international situation did not permit him to secure as extensive an ascendancy over surrounding regions as David had done; and both his own religious policy and that of his son and daughter-in-law aroused opposition which led to a religious crisis and ultimately to a politico-religious *coup d'état* in which his dynasty was overthrown.

The religious crisis did not come to a head until the reign of Omri's successor Ahab (869–850). No doubt this helps to

Foundation wall at Samaria.

account for the much greater space which is devoted in the Bible to the account of Ahab's reign (1 Kings 16²⁹–22⁴⁰); but it is not the sole explanation. Considerable attention is paid to Ahab's military achievements in repelling Syrian attacks. Syria (i.e. the Aramaean kingdom of Damascus) had been and continued to be a troublesome neighbour to Israel. Earlier in the century, when Baasha was threatening Judah's northern frontier, Asa had sent what remained of the Temple treasure to Ben-hadad, King of Damascus, as the price of Syrian intervention against Israel (1 Kings 15$^{17-22}$). The intervention had been to the military and commercial advantage of Syria and to the corresponding disadvantage of Israel; and Omri had sought to reinforce Israel's commercial position by his alliance with Phoenicia and its strategic position by the creation of a strong new capital. Syrian incursions into Israelite territory were a feature of the period; but in Ahab's reign Ben-hadad (possibly a second king of that name) attempted to subjugate Israel entirely. As a result of Ahab's spirited resistance, the Syrian army was driven back in disorder (1 Kings 20$^{1-21}$). The following year the Syrians struck again, only to be decisively defeated at Aphek, east of the Sea of Galilee (1 Kings 20$^{23-34}$); and as a result Ahab was able to regain territory and to secure trading advantages for Israel.

It appears that the impending danger from another quarter led Israel and Syria to sink their differences and to join in a defensive alliance with a number of neighbouring states. The armies of Shalmaneser III of Assyria (859–824) had already penetrated as far as the Mediterranean coast, but had withdrawn. In a second campaign in 853 they were met at Qarqar on the river Orontes by the forces of the western allies. Shalmaneser's own account of the battle (preserved on a stele in the British Museum) claims a crushing victory.[1] Since, however, he does not speak of any subsequent victorious advance, he probably exaggerated his achievement. The military strength of Ahab's kingdom is indicated by the size of the Israelite force, which included the largest contingent of chariots (2,000) in the army of the western allies. This battle is the first event in Israelite history which can be dated by the precise evidence of an extra-biblical document; but the Old Testament itself contains no record of it. For the next few years there was no further Assyrian invasion of the west. During this lull, hostilities again broke out between Israel and Syria. With the help of Jehoshaphat of Judah, Ahab attempted to recover Ramoth-gilead, which the Syrians still held; but he was killed in battle (1 Kings 22$^{1-40}$). The threat from Syria remained; but with

[1] See *ANE*, pp. 189–91; *DOTT*, p. 47.

Ahab's death the greatness of the Omrid dynasty was at an end.

The story of Ahab's last campaign recalls the internal religious conflict of his reign. He went out to war in spite of the ominous words of a solitary prophet of Yahweh whom he had imprisoned (1 Kings 22$^{7-28}$); and in the circumstances of his death there was held to be the fulfilment of a prophecy by Elijah (1 Kings 22^{38}; cf. 21^{19}). In contrast with the stories which depict him as a courageous and able military leader, there are others in which he appears as an apostate from pure Yahwism. In them, the commanding figure is not the king but the prophet Elijah; and even in the prosecution of royal policy the driving force is Ahab's Phoenician queen, Jezebel. The worship of Asherah and of Baal enjoyed royal patronage. The precise identity and nature of this Baal are disputed. It has been held that he was Baal Shamem (lord of heaven), and also that he was Melqart (king of the city, variously understood as king of the city of Tyre and as king of the underworld). At all events, the threat to the worship of Yahweh was serious. The queen's vigorous propagation of the Tyrian cult was a direct challenge to Yahweh's claim on the undivided allegiance of his people. Prophetic groups who withstood the royal policy were persecuted (1 Kings 18^4; 19^{10}). Nevertheless, the royal apostasy was publicly assailed by one man.

Elijah, a Gileadite from Transjordania, represents the most austere tradition of Hebrew prophecy. Though he had connexions with the prophetic communities (2 Kings 2$^{1-18}$), he was pre-eminently a solitary. He travelled alone or accompanied by a single attendant. At times he vanished mysteriously and then reappeared unheralded, with awe-inspiring suddenness. Unlike his successor Elisha, he was not an accessible or popular religious teacher; but he came to occupy in later tradition a position of peculiar importance.

The conflict in which he engaged concerned matters which were vital to the nation's faith. Other religions of the ancient Near East could tolerate the importation of alien deities and yet survive; but the acceptance of another god alongside

Yahweh was incompatible with the true nature of Israel's religion. When Elijah prophesied drought (1 Kings 17[1]), he was asserting that the power to give or withhold rain lay with Yahweh and not with Baal. When he challenged the prophets of Baal on Mount Carmel (1 Kings 18), he was again denying the ability of Baal to do what his worshippers claimed for him; and when he called upon the people to choose between Yahweh and Baal (1 Kings 18[21]), he was making plain Yahweh's claim on the undivided allegiance of Israel. His denunciation of the crime against Naboth (1 Kings 21) was an assertion of the ancient standards of communal justice in Israel. This combination of religious fidelity with social righteousness is characteristic of Old Testament religion, particularly in its prophetic expression. As the importation of Baal worship involved a denial of the supreme lordship of Yahweh in Israel, so the ruthless removal of Naboth and his sons was an outrage against the status and rights of the ordinary Israelite within the covenant community and against the sanctity of the administration of justice. Developments had already begun which were to be increasingly evident in the kingdoms of Israel and Judah: the pressure, not only of Canaanite fertility religion, but of the cults of foreign kingdoms; the threat to traditional norms of equity brought about by economic and social changes; and the corruption of judges and witnesses in the interests of the rich and powerful. Later prophets inveighed against these abuses. Their message was anticipated in Elijah's ministry, and not least in this, that he, like them, did not simply confront his hearers with abstract principles and standards to which they ought to conform, but spoke in the name of Yahweh, the living God, whose righteousness was a force to be reckoned with in history, who would not only rebuke Ahab, but bring judgement on him and his house (1 Kings 21[19 ff.]).

Ahab's successor, Ahaziah, reigned for only a short time (850–849). He was succeeded by his brother, Joram (849–842), in whose reign Moab reasserted its independence. We have both the Israelite and the Moabite accounts of this. 2 Kings 3[4–27] tells of a punitive campaign against Moab, in which

Joram was supported by Judah and Edom. The allies gained an initial success, but were later obliged to withdraw. The monument known as the Moabite stone[1] contains a record by the Moabite king Mesha of his country's subjugation by Omri, which Mesha attributes to the anger of the Moabite god Chemosh, and the subsequent achievement of independence, when Chemosh was favourable. There are some seeming discrepancies between the inscription and the biblical narrative; but the general similarity of background and of language is unmistakable. This military reverse is one indication of the decline of Israelite power.

The atmosphere of the period is effectively conveyed by the cycle of stories about Elijah's successor, Elisha (2 Kings 2–8), in which we catch glimpses of the court, the prophetic communities, and the life of the ordinary Israelite, made insecure by the incidence of drought and bad harvests, and by the periodic Syrian invasions. It was in the course of operations against the Syrians at Ramoth-gilead that opposition to the house of Omri reached a climax of violence (2 Kings 9 f). During Joram's absence from the camp, Elisha sent a prophet to anoint Jehu, one of the senior officers, as king. Assured of the support of his colleagues, Jehu immediately made for Jezreel, where he killed Joram. Ahaziah, king of Judah, who was in Jezreel at the time, and Jezebel were also put to death. Before proceeding to Samaria, Jehu ordered that all members of the former royal family and of the court should be killed; *en route* there he slaughtered a visiting party of Judaean princes; after his arrival he treacherously massacred the Baal worshippers. Whatever elements of personal ambition and political expediency may have prompted Jehu's actions, he had the support of the religious conservatives, notably Jonadab the Rechabite (2 Kings 10[15 f.]), whose clan remained doggedly faithful to the ways of semi-nomadic life. But a century later one of the greatest of the prophets saw in Jehu's actions an outrage which called down the judgement of Yahweh (Hos. 1[4]).

[1] See *ANE*, pp. 209 f.; *DOTT*, pp. 195–8.

Inscription of Mesha.

During the entire duration of the Omrid dynasty there was a particularly close association between the kingdoms of Israel and Judah and also between their royal houses. Asa's successor, Jehoshaphat (873–849), was an active ally of Ahab, whose daughter, Athaliah, was married to Jehoshaphat's son, Jehoram (2 Kings 8¹⁸). Jehoshaphat's unsuccessful attempt to re-establish sea-borne trade based on Ezion-geber (1 Kings 22⁴⁸) is an indication that Judah exercised some degree of control over Edom, through whose territory the route to Ezion-geber ran. Jehoshaphat also extended his influence to Philistia and Northern Arabia (2 Chron. 17¹¹). But during the reign of his son Jehoram (849–842) there were successful revolts in both Edom and Philistia (2 Kings 8²²). Important developments also took place in the religious field. Jehoshaphat had carried out certain judicial reforms (2 Chron. 19⁴⁻¹¹); and had been a faithful worshipper of Yahweh. Jehoram's religious policy was very different. Under the influence of Athaliah, he established the Baal cult in Judah (2 Kings 8¹⁸); and when, after a reign of only one year (842), Jehoram's son, Ahaziah, was killed in Jehu's revolt, Athaliah seized power and put to death all the members of the royal family except Ahaziah's infant son, Joash, who was rescued and hidden by his aunt, the wife of the chief priest Jehoiada (2 Kings 11¹⁻³). For six years Athaliah ruled the land. Then, in a swift and completely successful revolt, she was assassinated and the boy Joash was acclaimed as king (2 Kings 11⁴⁻²⁰). This Yahwistic rising in Judah differed from the northern one in three ways: it was engineered, not by a prophet, but by a priest; there does not appear to have been a general massacre of Baal worshippers (possibly because the Baal cult had enjoyed less popular support in Judah than in the north); and the revolution was achieved, not by the destruction of the royal line but by its restoration. In the event, Joash (837–800) proved himself to be the kind of king who could win the approval of the Deuteronomistic historian, who records at some length his care for the Temple (2 Kings 12¹⁻¹⁶).

But difficult days lay ahead for both kingdoms. Jehu (842–815), whose violent methods of gaining the throne might be thought to indicate energy and initiative, failed to restore Israel's strength and made no great showing in his relations with other countries. The king of Syria had been murdered and supplanted by Hazael (2 Kings 8[7-15]), who became a troublesome adversary of both Israel and Judah. The Syrians themselves had to face renewed attacks by Assyria, notably in 841, when Shalmaneser III besieged Damascus but failed to take it. Jehu was one of a number of western rulers who, in the course of the same campaign, were compelled to pay tribute, as recorded on the Black Obelisk of Shalmaneser.[1] Assyria's preoccupations elsewhere enabled Hazael to occupy Israelite territory east of Jordan (2 Kings 10[32f.]), and later to impose on Jehu's son, Jehoahaz (815–801), a humiliating reduction of his armed forces (2 Kings 13[7]). The Syrians penetrated as far as the southern coastal plain, captured Gath, and threatened Jerusalem, but were bought off by Joash (2 Kings 12[18]).

The position of the Israelite states was eased by contrasted phases in the activities of Assyria. Towards the end of the century, Adad-Nirari III carried out a series of campaigns in the west, in the last of which he broke the power of Syria. But later the Assyrians themselves were so hampered by internal weakness and external dangers that they were unable to intervene effectively in the west. Thus the Israelite kingdoms were temporarily freed from the incursions of their northern neighbour and also from Assyrian pressure. The way was open for a partial renewal. Jehoash of Israel (801–786) regained the territory which had been lost to Syria (2 Kings 13[25]). Amaziah of Judah (800–783) who succeeded to the throne after the assassination of his father Joash (2 Kings 12[20f.]), was able to subjugate Edom (2 Kings 14[7]). But when he rashly involved himself in a quarrel with Israel, his army was defeated, his capital stormed and plundered, and he himself taken prisoner (2 Kings 14[8-14]; cf. 2 Chron. 25[5-24]). His subsequent fortunes are obscure, except that, like his father, he was assassinated (2 Kings 14[19f.]).

[1] See *ANE*, p. 192; *DOTT*, p. 48.

Black Obelisk of Shalmaneser (Jehu's Tribute).

*From the Reigns of Jeroboam II of Israel and Uzziah of Judah
to the Death of Hezekiah of Judah (786–687 B.C.).*

There now began a period of nearly half a century in which
both Hebrew kingdoms enjoyed a stability and prosperity
which recalled the great days of the undivided kingdom. They
were free from serious external interference; they were able to
recover lost territories; they had scope to develop trade; and
in each of them an able king reigned for some forty years;
in Israel, Jeroboam II (786–746); and in Judah, Uzziah or
Azariah (783–742).

Taking advantage of continuing Syrian weakness, Jeroboam
pushed Israel's frontier north as far as the approach to Hamath
and occupied considerable territories east of Jordan (2 Kings
14²⁵). In an invasion of Philistia, Uzziah captured Gath,
Ashdod, and Jabneh (2 Chron. 26⁶). Following up his father's
conquest of Edom, he rebuilt the port of Elath at the head of
the Red Sea (2 Kings 14²²), thus securing lines of communication which were important for trade. The commercial importance of the territorial gains made by both kings was the
greater because the kingdoms were on good terms with each
other. Peaceful relations made for easy communication,
prosperity, and a renewed sense of national confidence.

But in spite of the restoration of wealth and strength, there
were serious weaknesses in both kingdoms, and particularly in
Israel. The teaching of the eighth-century prophets reflects
vividly the social disintegration and corruption of the period.
The ancient structure of Israelite society had been seriously
affected by economic and other developments which had taken
place at least since the establishment of the monarchy. The
centralization of administration had undermined local independence. The internal Israelite economy, which was always
liable to be affected by drought and bad harvests, had suffered
seriously because of the prolonged wars with Syria and other
neighbouring countries. The combination of adverse factors
bore heavily upon the small farmer. In hard times he was
driven to borrow; and, when he could not pay his debts, the

mortgage on his land might be foreclosed and his property for-
feited. From being a freeman, he was deprived of his ancestral
land and reduced to the position of a labourer. The plight
of the poorer members of the community was made the
more miserable because of the widespread corruption of the
administration of justice: the rich were not slow to take advantage of the venality of judges and witnesses. Thus, in place of the
ancient principle of brotherhood in Israelite society, there had
arisen a new order in which a wide gulf separated the rich
from the poor. The ivory inlays from this period discovered at
Samaria have illustrated the prophetic denunciations of the
luxury in which the rich lived.[1]

These internal weaknesses were soon to be matched by
dangers from abroad. The relative peace of the first half of the
eighth century presents a striking contrast to the series or
crises which had to be faced in its closing decades. The last
and greatest period of Assyrian expansion was inaugurated by
the usurpation of power by Tiglath-pileser III (745–727).[2]
Almost at once he embarked on a series of campaigns in the
west. These were followed, during the reigns of his successors,
by others in which Assyria destroyed or gained control of the
small states in the west, and finally, in the following century,
carried out a successful invasion of Egypt itself.

In Judah, Uzziah was unable because of illness (2 Kings 15[5])
to retain personal control of the government during the closing
years of his life. His son Jotham acted as regent, and later
reigned as king in his own right (742–735) just at a time when
external pressures were beginning to threaten. But the major
crises of the period developed during the reigns of Ahaz
(735–715) and Hezekiah (715–687). In the northern kingdom
the external dangers were matched by a series of dynastic
crises. Zechariah, the son and successor of Jeroboam II, had

[1] Among the biblical passages relevant to the conditions described above,
the following may be specially noted: Amos 2[6–8], 5[11 f.], 6[4–6]; Isa. 5[8–10], 22[f.];
Mic. 2[1 f.]. These show that the abuses were prevalent in both kingdoms.

[2] This king was also known by his Babylonian name 'Pul'; see, e.g.,
2 Kings 15[19].

reigned for only six months when he was murdered and supplanted by Shallum. A month later, Shallum was dethroned by Menahem (745–738), during whose reign Israel, like her northern neighbour Damascus, became tributary to Assyria. Menahem's son Pekahiah was murdered after a reign of some two years (738–737), perhaps because there were those who wanted the country to retain its independence. At all events, Pekahiah's assassin and successor, Pekah (737–732), combined with the Syrian king Rezin (Rezon) in an alliance designed to present an effective resistance to Assyria. The two northern powers attempted to induce or coerce Judah into joining the coalition. Judah's refusal led, shortly before Jotham's death (2 Kings 15³⁷), to an invasion of her territory which continued after the accession of Ahaz (2 Kings 16⁵; cf. Isa. 7¹⁻⁹). Clearly the situation was serious. Jerusalem itself was besieged; and one of the aims of the Syro-Ephraimite confederacy was to depose Ahaz and replace him by an Aramaean who is referred to as 'the son of Tabe-el' (Isa. 7⁶). In other quarters Judah was hard pressed. Elath was lost;[1] and the Philistines overran parts of the Negeb and the Shephelah (2 Chron. 28¹⁸). When the young king Ahaz turned to Assyria for help against his northern neighbours, the prophet Isaiah urged him to trust solely in Yahweh's protection (Isa. 7³ff·). It is perhaps not surprising that Ahaz persisted in his intention; but his appeal was in fact unnecessary. Tiglath-pileser would have been obliged, in any event, to take measures against the confederacy. He did so with remarkable effectiveness, striking as far south as Philistia, then turning north against Israel out of whose territory he carved three new Assyrian provinces, Gilead, Megiddo, and Dor (2 Kings 15²⁹). Finally, in 732, he captured Damascus, executed Rezin, and divided the Syrian kingdom into four provinces (2 Kings 16⁹). In Israel, these events were accompanied by yet another dynastic *coup d'état*. Pekah was assassinated by Hoshea, who then ruled the rump kingdom

---

[1] It is uncertain whether this was brought about by Syrian intervention or by an Edomite revolt. See 2 Kings 16⁶, and cf. R.S.V. note. The Hebrew words for Syria (Aram) and Edom look alike.

(732–724) as a tributary vassal of Assyria (2 Kings 15³⁰).[1] Whether or not the Chronicler is right in stating that Judah also suffered in this campaign (2 Chron. 28²⁰), Ahaz had to pay heavy tribute to Tiglath-pileser. There were also religious consequences of his political submission. Various changes were made in the Temple, notably the erection of an altar modelled on one which Ahaz had seen when he went to do homage to Tiglath-pileser in Damascus (2 Kings 16¹⁰⁻¹⁸). But quite apart from this, Ahaz actively encouraged pagan practices and even offered his son as a sacrifice (2 Kings 16²⁻⁴). His reign was a gloomy time of territorial and financial losses, political subservience, and apostasy.

The northern kingdom did not long survive the fall of Damascus. When Shalmaneser V succeeded Tiglath-pileser, Hoshea at first paid tribute to him (2 Kings 17³); but soon afterwards he entered into negotiations with So (Sib'e), an Egyptian local king or commander, and withheld his tribute from Assyria. In consequence, he was arrested, the country was overrun by the Assyrians, and Samaria itself was besieged (2 Kings 17⁴ f.). Three years later, after the death of Shalmaneser, the city was taken by his successor Sargon II, who deported part of the population to Mesopotamia and Media and brought into Israelite territory settlers from other parts of the Assyrian empire (2 Kings 17⁶, ²⁴).[2]

The destruction of the northern kingdom, important as it must have seemed in the perspective of Israelite history, was only an incident in the Assyrian attempt to keep the west in subjection and to foil Egyptian schemes to foment rebellion among Assyria's vassals. For the surviving state of Judah it was a recurring question whether to remain loyal to her Assyrian overlord or to become involved in projects of revolt. The policy of Ahaz had been one of submission. During the reign of his son Hezekiah (715–686) important changes took place both internally and externally. Reversing his father's

---

[1] See *ANE*, p. 194; *DOTT*, p. 55.
[2] For Sargon's accounts of his operations against Samaria see *ANE*, pp. 195 f.; *DOTT*, pp. 59–61.

religious measures, Hezekiah carried out a programme of cultic reform (2 Kings 18⁴), in which he abolished not only the pagan practices which Ahaz had introduced but also others of much greater antiquity. He also attempted to suppress the high places, and, according to the Chronicler (2 Chron. 30¹⁻¹¹), invited the northern Israelites to join in worship at Jerusalem. In its double aim of purification and centralization Hezekiah's reform resembles the one carried out by Josiah nearly a century later. There is another similarity. Both were gestures of political independence. The elimination of Assyrian religious influence implied the repudiation of Assyrian political authority. In the field of foreign policy one of the most important factors was the revival of Egyptian power under the Twenty-fifth (Ethiopian) Dynasty, which was established c. 716. Assyrian expansion in the west was a threat to Egypt. In self-defence, Egypt tried to use Judah and her neighbours as buffer states, encouraging them to revolt, promising help, and, on occasion, sending her armies north. During the closing years of the century there was intense diplomatic activity between Egypt and Judah. The prophet Isaiah, who is commonly credited with having helped to inspire Hezekiah's religious reforms, was resolutely opposed to this aspect of national policy, as he was to all such foreign entanglements. Several of his prophecies reflect vividly the coming and going of envoys and predict the disastrous outcome of the intrigues (e.g. Isa. 18¹⁻⁷, 20, 30¹⁻¹⁸, 31¹⁻³). For a time Judah remained uncommitted. A revolt led by the Philistine city of Ashdod in 713 resulted in an effective punitive campaign by the Assyrian armies two years later (Isa. 20¹). The rebellious ruler of Ashdod sought asylum in Egypt, but was handed over to the Assyrians. Since Judah was left unscathed, it may be presumed that Hezekiah had not been actively involved in the revolt.[1] But a few years later he joined in an alliance against Assyria which had extensive ramifications.

In 705 Sargon was killed in battle. His successor, Sennacherib, was immediately confronted by disturbances in both

[1] See *ANE*, pp. 197 f.; *DOTT*, pp. 61 f.

east and west. A Chaldaean chieftain called Marduk-apal-iddina, who had established himself as king of Babylon in 720, but had been deposed by Sargon in 709, revolted and seized power again in Babylon. This is the Merodach-baladan (or Berodach-baladan) of the Bible (2 Kings 20¹²⁻¹⁵; Isa. 39¹⁻⁴), who sent envoys to Jerusalem to encourage resistance to Assyria; and though it is not impossible that these overtures were made at an earlier date, it seems most likely that they should be dated shortly after Sennacherib's accession, and that their aim was to ensure that Assyria would have to fight on two fronts. The revolt in the west was backed by promises of Egyptian help. A number of Phoenician cities, and, in Philistia, Ashkelon and Ekron, were all actively involved. Padi, the pro-Assyrian ruler of Ekron, was deposed by his subjects and handed over to Hezekiah for safe keeping. But the plans to break Assyria's imperial power failed miserably. Sennacherib's first move was to recapture Babylon and expel Marduk-apal-iddina. In 701 he turned west, subdued Phoenicia, and marched south against the remaining rebels. An Egyptian army which had advanced to relieve Ekron was defeated at Eltekeh. The territory of Judah was ravaged; and Hezekiah was obliged to release Padi and to pay heavy tribute; but Jerusalem did not fall. Sennacherib's own lively account of the campaign has been preserved on the Taylor Prism, now in the British Museum.[1] The Old Testament evidence presents some difficulties. (a) In 2 Kings 18¹³⁻¹⁶ there is a summary account of Sennacherib's invasion and Hezekiah's submission and payment of tribute. (b) 2 Kings 18¹⁷⁻19⁸ tells how Sennacherib sent from Lachish a considerable force demanding the surrender of Jerusalem. Hezekiah was assured by Isaiah that Sennacherib would receive news which would oblige him to return to Assyria. The Assyrian threat was not pressed; and before long the force rejoined the main army, now at Libnah. (c) 2 Kings 19⁹⁻³⁷ describes how, when Sennacherib heard that Tirhakah, king of Ethiopia, was advancing against him, he sent messengers to Hezekiah with a threatening letter. Again

[1] See *ANE*, pp. 199–201; *DOTT*, pp. 66 f.

Sennacherib's attack on Lachish.

Defenders of Lachish.

Isaiah assured his king that Yahweh would protect the city. The same night 185,000 men in the Assyrian army were killed by the angel of Yahweh. Sennacherib withdrew the remainder of his force to Assyria, where he was assassinated by two of his sons.[1] It can be argued that (b) and (c) are two variant accounts of the same events, or that they describe two successive phases in the one campaign, or that (b) refers to the invasion of 701 and (c) to a later invasion assumed to have taken place in 698 or 687 (of which, however, there is no Assyrian record). The evidence is complex; but it seems best to adopt the second of these views. At first, we may suppose, Sennacherib sent a strong force against Jerusalem, but was unable to make Hezekiah surrender the city. Later, hearing of the approach of an army from Egypt, and being unable to spare a substantial number of men, he sent a threatening letter to Hezekiah, hoping to cow him into submission. The calamity which befell the Assyrian army is usually presumed to have been a sudden pestilence, possibly bubonic plague. Of this the Assyrian record (not surprisingly) says nothing; but Herodotus records (*History* 2, 141) an Egyptian tradition that Sennacherib's army had to retreat because during the night mice gnawed the quivers, bowstrings, and shield-handles. Since mice and rats spread bubonic plague, this story may be a distorted record of such an outbreak of disease. The reference in the Old Testament to 'Tirhakah king of Ethiopia' (2 Kings 19⁹) presents a difficulty. Tirhakah did not become king until some years later, and in 701 was probably only 9 or 10 years old. The occurrence of his name may well be an accidental anachronism. It should also be noted that the assassination of Sennacherib, which is mentioned at the end of the Old Testament narratives of the campaign, in fact took place in 681.

Though Jerusalem had survived, Judah had suffered terribly and had been left in a pitiful plight. Isaiah, who describes vividly the jubilation in the capital when the Assyrians withdrew (Isa. 22¹⁻⁴), also presents in grim terms the devastation

[1] 2 Kings 18¹³, ¹⁷–19³⁷ is repeated, with some variations, in Isa. 36–37. The account in 2 Chron. 32¹⁻²³ is somewhat briefer.

The Siloam Tunnel.

which had been wrought (Isa. 1$^{5-9}$). There could be no quick recovery from such a calamity. For the greater part of the following century Judah remained under the effective domination of Assyria.

Apart from the narratives of the events, one striking reminder of the crisis still survives. This is the tunnel which Hezekiah's workmen had hewn from the spring Gihon (known today as the Virgin's Fountain) to the pool of Siloam inside the city walls to provide a better water-supply in the event of siege (2 Kings 20^{20}).[1]

It is from this period of internal degeneration and external crisis that there have come down to us the earliest considerable records of prophetic teaching. In the stories about the prophets of earlier ages (e.g. in the Elijah cycle) there are occasional brief utterances of judgement and the like. But these are not really comparable with the compilations of teaching in the books of the prophets (Isaiah, Jeremiah, Ezekiel, and the twelve Minor Prophets). The men by whose names these books are known have sometimes been called 'writing prophets'; but this term is misleading, not because these prophets could not or did not on occasion write, but because most of them communicated their teaching in the first instance by speech and action. Only subsequently was it preserved in writing by themselves or by others. The prophetic books do not, for the most part, contain sustained discourses, but are compilations of prophetic utterances (often in poetical form), and of biographical and autobiographical narratives, put together without strict regard for logical and chronological sequence, preserved, transmitted, and augmented in the main by the prophetic disciples who were the heirs of the great masters.

The activity of the prophets whose teaching has been thus preserved spanned a period from the middle of the eighth century until well on in the post-exilic age. Our immediate concern is with those who lived and worked during the latter

[1] The execution of this considerable feat of engineering is described in a contemporary inscription found in the tunnel in 1880 and now preserved in a museum in Istanbul. For the text see *ANE*, p. 212; *DOTT*, p. 210.

Siloam inscription.

part of the eighth century: Amos, Hosea, Isaiah, and Micah. These men have sometimes been regarded as representatives of a wholly new type of prophecy and even as the founders of Israel's distinctive religion. But this is to underestimate their debt to the past and their links with earlier prophecy. They came forward as the spokesmen of the God who had delivered Israel from Egyptian bondage and given her the land of Canaan for her inheritance; and they sought to recall Israel to the traditional standards of covenant faithfulness to God and man. In this last point there is a close and obvious link between them and Elijah in the previous century; and there are other features which they shared with the prophetic movement as a whole: the name *nabhi*, abnormal psychological experiences (though the evidence for such experiences is much less frequent in the canonical prophets, with the exception of Ezekiel, than is sometimes supposed), and the use of dramatic symbolism. They stand in sharp contrast, not so much to an earlier phase of prophecy as to features in the prophetic movement which existed before, during, and after their time: a professionalism and nationalism which obstructed the true perception and faithful proclamation of the will of Yahweh. This contrast is made explicit in the declaration by Amos that he prophesied, not because he had belonged to a community of professional prophets, but because he had been personally called to the task by Yahweh (Amos 7¹⁴ ᶠ·). The same distinction appears, over a century later, in Jeremiah's denunciation of the prophets who purveyed oracles but had not stood in the council of Yahweh (Jer. 23⁹⁻²², especially v. 18). These two instances make plain the deep sense of vocation and the intense preoccupation with the will of Yahweh which characterized the great prophets. In this they represented a prophetic tradition in Old Testament religion which cannot be equated with the entire prophetic movement as such, and which went back to the Mosaic formulation of Israel's faith, if not, indeed, to the religion of the patriarchs.

The nature of their preoccupation with the will of Yahweh makes it clear that the great prophets were not mere traditionalists. Their references to what Yahweh had done for Israel in the past were closely related to what they had to proclaim about the meaning of contemporary and future events. They were the spokesmen of a God who not only had acted in the days of the Exodus and the settlement, but was acting now in the history of Israel and of other nations. Their condemnation of the corruption of society in their own day was not an attempt to put back the hands of the clock and make the whole nation return, like the Rechabites, to the social and cultural conditions of an earlier age. It was, rather, a claim that the ancient standards of justice, brotherhood, and compassion were still binding in the drastically altered conditions of their own time.

The appeal for justice is expressed with particular emphasis in the teaching of Amos, the earliest of the canonical prophets. He is described as 'among the shepherds of Tekoa' (Amos 1^1). Evidence from Ugarit and elsewhere shows that the word translated 'shepherd' could be used of a temple functionary; but it was not necessarily so used and there is no reason to suppose that Amos had any special cultic status. He prophesied because Yahweh had called him to do so and had given him a word to speak (Amos. $3^{7 f.}$, $7^{14 f.}$). Though his home was at Tekoa in Judah, it was to the royal sanctuary at Bethel that he went to prophesy, and it was Ephraimite society that he castigated. The few allusions to Jerusalem and Judah in the book seem to be either incidental to the prophet's main message or later additions. The time was the middle of the eighth century. Jeroboam II was still on the throne (Amos 1^1, 7^{10}); and Israel's period of security and prosperity had not yet ended. Amos foretold the coming humiliation of Israel (5^{27}, $7^{9, 11, 17}$); but there are no indications in his recorded prophecies that the blows had already begun to fall. The confidence and complacency of the nation were still unshaken.

In trenchant terms Amos denounced the selfish luxury of the rich, their callous oppression of the poor, the dishonesty of merchants, and the corruption of the law courts (2^{6-8}, 3^{10}, 4^1, 6^{4-6}, 8^{4-6}). Such abuses were made all the more heinous because they were accompanied by the enthusiastic observance of the cult at the great sanctuaries (2^8, $4^{4\,f.}$, $5^{5,\,21-23}$); the outwardly religious nation had failed to hear and obey Yahweh's demand for righteousness and justice (5^{24}).

Criticism of the corruption of Israel's religion by alien elements does not seem to have been a prominent feature in the prophet's teaching (5^{26}, 8^{14}); but it would be a mistake to suppose that he was simply a social reformer. His condemnation of social injustice in Israel presupposed the story of the Exodus and the settlement in Canaan. The people had not only broken the laws of right conduct: they had flouted the commands of the God to whom they owed their life as a nation, the land in which they lived, and their continuing religious tradition (2^{9-12}). The point is emphasized by the way in which the exposure of Israel's sin comes as the climax of a series of denunciations of other peoples (1^3-2^3), all of whom had committed atrocities which were acts of rebellion[1] against Yahweh and which would be punished by Him. There is no suggestion that these other nations had a special responsibility to the gods whom they worshipped rather than to Yahweh. They were subject to the will of Yahweh and liable to punishment by Him, even when they had done wrong to a nation other than Israel (2^1). This idea (which it seems that Amos could assume in the minds of his hearers), that other nations were morally responsible to Yahweh, is of great importance for the understanding of the Old Testament belief in Yahweh as the one true God. Israel's responsibility, however, was the more serious because of her special relationship to Yahweh: 'Hear this word that the LORD has spoken against you, O people of Israel, against the whole family which I brought up out of the land of Egypt: "You only have I known of all the families of the earth; therefore I will punish you for all your iniquities"' ($3^{1\,f.}$). This statement implies election and covenant, though the actual terms are not used. Complementary to it is another which makes it clear that Yahweh retains His sovereign free-

[1] This is the meaning of the word which is translated 'transgressions' in A.V., R.V., R.S.V.

dom, and is not bound exclusively to any one nation but is master of all nations: ' "Are you not like the Ethiopians to me, O people of Israel?" says the LORD. "Did I not bring up Israel from the land of Egypt, and the Philistines from Caphtor, and the Syrians from Kir?" ' (9⁷).

Amos's prediction of judgement on Israel is most pointedly put in his words about the day of Yahweh (5¹⁸). His hearers believed that this time of Yahweh's decisive action would bring national prosperity and triumph; but Amos declared that it would be a day of affliction and loss. The term 'the day of Yahweh' may be drawn from warfare (meaning the day of Yahweh's victory) or from the nomenclature of the autumnal festival (meaning the day when Yahweh's kingship is manifested). It had a long history in later prophetic and apocalyptic teaching and came ultimately to denote the final crisis in which Yahweh would inaugurate a wholly new order. For Amos it meant the reversal in the immediate future of the people's hopes. He is pre-eminently the prophet of doom. The glowing picture of restoration with which his book ends is commonly regarded as a later addition. Elsewhere in his oracles there is scarcely a gleam of hope (5⁴, ⁶, ¹⁴). His teaching thus raises in an acute form a question which is also suggested by some of the prophecies of doom in other pre-exilic prophets: Do such prophecies imply the annulment of the covenant between Yahweh and Israel and the total and final destruction of the national life?[1] Some scholars have answered this question in the affirmative so confidently that they have been obliged to treat as later additions those oracles of promise and hope which appear in the books of the pre-exilic prophets. It is clear that predictions of the destruction of the *northern* kingdom do not imply the end of Yahweh's purpose for His people, since Judah survived as heir of the name Israel. But there are other considerations which apply to predictions of doom on either kingdom. The prophets were *prophets*, not logicians or systematic

---

[1] The word 'covenant' is curiously absent from the teaching of the canonical prophets before Jeremiah (cf. Hos. 6⁷, 8¹); but the absence of the word does not necessarily imply the absence of the idea.

theologians: in any given situation they were concerned to drive home a particular truth as forcefully as they could, and not to qualify or modify their expression of it in order to relate it to what they had to say in other situations. Further, the fact that the prophets summoned the people to repentance suggests the possibility either that disaster might be averted or that beyond what looked like irretrievable calamity there still lay a purpose of Yahweh. At all events, the prophets make it clear that neither Israel's independent existence nor her material prosperity are ends in themselves, and that Yahweh's special relationship to Israel is only part of a divine purpose in which other nations are involved. It is in this context that we must understand the protests of Amos against the corruption of Israelite society. When he denounces the unrighteous action of men he is also the herald of the righteous action of Yahweh. In this he must be regarded as among the greatest of the prophets.

Hosea's prophecies, like those of Amos, are addressed to the northern kingdom; but they come from a slightly later time and reflect important changes in the national situation. Allusions to internal dynastic crises and to vacillating foreign policy (7^{7-11}, 8^4) suggest the period of political instability which preceded the fall of Samaria.

There are, in fact, two backgrounds to Hosea's teaching: the national life and his own domestic life. Chapter 1 tells part of the story of the marriage of Hosea to Gomer and of her infidelity; chapter 3 seems to return to the same theme. Much scholarly ingenuity and some ill-judged romantic fantasy have been devoted to attempts to reconstruct the sequence of events. Are chapters 1 and 3 successive acts in the drama, or parallel accounts of the same events? Do these chapters refer to two different women, or is Gomer the woman mentioned in chapter 3? Was Gomer a temple prostitute? Did Hosea knowingly or unknowingly marry an immoral woman, or was his wife unfaithful to him only after they had been married for some time? We may note these questions; but we need not pause to discuss them. Hosea's primary purpose was not to describe his domestic tragedy but to proclaim the word of Yahweh to His people. It is, therefore, not surprising that he has not provided us with adequate autobiographical details. On the whole, the most likely view is that Hosea became aware of his wife's immorality after they had been married for some time, probably after the birth of the first child. Hosea believed that what had befallen him was of divine appointment (1^2) and saw in his tragic experience a counterpart to the relations between Yahweh and Israel (2, 3): his marriage corresponded to the establishment of the covenant bond, Gomer's infidelity to the apostasy of Israel, the estrangement between husband and wife to Yahweh's punitive discipline of His people, and Hosea's enduring love for his faithless wife to Yahweh's steadfast purpose of good to Israel. In 11^{1-11} the same pattern is presented in terms of the relationship between a father and his child.

Like Amos, Hosea presupposes the traditions of the Exodus and the settlement (2^{15}, 11^1); and it is against that background that he formulates his arraignment of the nation. Much of his polemic is directed against the adoption in Israelite religion of Canaanite fertility cult practices including the use of bull images ($8^{5\,f.}$, 10^5). Underlying the fertility cult was the assumption that the produce of the land came from the Baals (2^5): Hosea declares that it is Yahweh who has the power to give or withhold ($2^{8,\ 12,\ 21\,f.}$). In thus asserting that the whole realm of nature comes within Yahweh's sphere of operation, Hosea is in some sense borrowing from Canaanite religion even while he seeks passionately to combat it. The same may be said of his use of the symbolism of love and marriage; for the sacred marriage was an element of prime importance in the kind of worship which he denounced as apostate. Hosea was no Rechabite, rejecting all that belonged to agricultural society as incompatible with the religion of Yahweh. Yet he was aware of the danger of remaining nominally faithful to Yahweh but worshipping Him as if He were a Baal (2^{16}; cf. above, p. 67). In this he was true to the historical character of Israel's faith.

Hosea depicts the condition of Israelite society in gloomy colours: widespread immorality and corruption, political insecurity and violence, depravity and irresponsibility in the ranks of the priesthood. In $4^{1\,f\cdot}$ the essence of Yahweh's contention against His people is tersely presented: the abuses of the time come from the absence of 'faithfulness', 'kindness', and 'knowledge of God'. All three terms are important; but the second is held to be particularly characteristic of Hosea's teaching. In Amos, it is said, the emphasis is on righteousness and justice; but in Hosea mercy is central. This is misleading. The Hebrew word *hesedh*, which is here rendered 'mercy' (A.V. and R.V.) or 'kindness' (R.S.V.) expresses a combination of duty and affection, and is more satisfactorily translated by 'steadfast love' or 'devotion'. It is closely related to the covenant idea, and is therefore appropriate to Hosea's main charge against the people and his use of marriage symbolism. But, since 'righteousness' and 'judgement' are words with covenantal associations, and since *hesedh* implies faithfulness as well as feeling, there is no real contrast between the two prophets when, in denouncing Israel's worship, Amos says, 'Let justice roll down like waters, and righteousness like an ever flowing stream' (5^{24}), and Hosea says, 'I desire steadfast love (*hesedh*) and not sacrifice' (6^6).

Nor should the word 'mercy' mislead us into overlooking or underestimating the element of doom in Hosea's teaching. This is symbolically expressed in the names given to Gomer's three children ($1^{4,\ 6,\ 9}$) and recurs in varying ways throughout the book. But beyond chastisement there lies the possibility of restoration ($2^{14\,f.,\ 18-23}$; cf. $11^{8\,f.}$).

Micah and Isaiah both belonged to Judah; and in their prophecies Judah is the main object of denunciation. They are said to have prophesied during the last four decades of the eighth century (Isa. 1^1; Mic. 1^1). In their respective backgrounds and in the range of their teaching they differ markedly from each other.

Micah was a countryman from the small town of Moreshethgath in the Shephelah. His prophecies reflect the general conditions rather than the specific events of the time. He condemned the heartless oppression of the poor by the rich (2^{1-5}, $8^{f.}$, 3^{1-4}, $9^{f.}$) and the venality of the professional prophets and of the priests (3^{5}, ^{11}). Such flagrant disregard of righteousness and justice would lead to devastating punishment: both Samaria and Jerusalem would be destroyed ($1^{5 f.}$, 3^{12}). This much may be learned from 1–3, which is commonly regarded as the only part of the book in which Micah's own teaching is recorded. It leaves us with the picture of a prophet whose abhorrence of injustice and predictions of impending doom recall the teaching of Amos. The analysis and the dating of 4–7 raise difficult problems. Though some of the material probably comes from Micah's prophetic successors, there are other passages which may well go back to his own time and even to Micah himself. There is no conclusive argument for denying to him 6^{1-8}, a passage which sums up the prophetic indictment of Israel's sin as ingratitude to Yahweh, the Saviour God. Its closing verse states in classic simplicity what Yahweh requires of man: the justice for which Amos pleaded, the steadfast love (*hesedh*) which is central to the teaching of Hosea, and the humble trust which Isaiah saw as the only fit response of man to the holiness of Yahweh.

The work and teaching of Isaiah are recorded in chapters 1–39 of the book which bears his name. Even in these chapters there are some passages which come from considerably later periods (notably 24–27 and 34–35); but there is enough to enable us to reconstruct the outlines of a prophetic ministry which lasted for a full generation, from the end of Uzziah's reign until at least the turn of the century.

Isaiah was a citizen of Jerusalem. The common assertion that he was of aristocratic or even royal descent is without secure foundation. Whether or not he belonged to the court circle, it is as a prophet that he must be understood. His teaching reflects a mind of commanding range and power, able to subject contemporary conditions and national policy to penetrating scrutiny. We have already noted his interventions in three critical situations: his appeal for faith in Yahweh at the time of the Syro-Ephraimite invasion of Judah, his protests against negotiations with the Ethiopian rulers of Egypt, and his assurance that Jerusalem would not fall to Sennacherib in 701. We must now look at his teaching as a whole.

The account of his call (6) is of fundamental importance. His vision, in the Temple, of Yahweh seated on His throne brought home to him the awful holiness of Yahweh. Though Isaiah's first response was a devastating sense of guilt and unworthiness, holiness is not to be understood simply in the sense of ethical righteousness (cf. above, p. 35). It is, rather, the transcendent majesty of the living God, who is other than man and Lord of the whole world. Failure to acknowledge the holiness of Yahweh results on the one hand in pride, self-sufficiency, and reliance on wealth, arms, and alliances (see especially 2^{6-22}, $30^{1 \text{ f.}}$, 31^{1-3}), and, on the other, in panic in time of danger (7^{1-9}). The true response to 'the Holy One of Israel' is humble trust. It was to this that Isaiah tried to recall the people and their rulers during the repeated crises of his time.

Like his predecessors Amos and Hosea, and his younger contemporary Micah, he attacked social corruption and the oppression of the poor. The luxury-loving women of Jerusalem, the land-grabbers, the drunkards, the sceptics, the scoffers at morality, the self-sufficient, the perverters of justice, all come under the lash of his condemnation (3^{16}–4^{1}, 5^{8-23}). But, as with Amos and Hosea, the passion for social righteousness is related to the goodness of Yahweh to Israel and the special bond which was established thereby. This is expressed in the Song of the Vineyard (5^{1-7}), which may have been chanted or sung by Isaiah on the occasion of a vintage festival, using the setting and the symbolism of agricultural religion to expose Israel's faithlessness.[1] For such faithlessness and cruelty no prodigality of sacrifice or zealous religiosity could atone (1^{10-17}). The coming divine punishment for the nation's sin is described in three main ways: in terms of natural disasters

[1] It is probable that the language used about the vineyard also symbolizes the marriage relationship (Song of Sol. $1^{6, 14}$, 2^{15}, 8^{12}).

on the day of Yahweh (2^{6-22}), as a process of internal disintegration in the life of the community (3^{1-15}), and as inflicted by the armies of Assyria (10^{5-34}). The northern kingdom, as well as Judah, would suffer Yahweh's chastisement. This is predicted, not only in connexion with the invasion at the beginning of the reign of Ahaz (7^{16}), but in a prophecy about the final fall of Samaria (28^{1-4}). The reference to Assyria as the rod of Yahweh's anger (10^5) is of great importance. To punish His people and to further His purpose Yahweh could use even a tyrannous power seeking its own aggrandisement.

Isaiah's teaching also contains the positive element of promise. It is a mistake, however, to transform his predictions that Jerusalem would not be captured into a doctrine of 'the inviolability of Zion'. These predictions referred to specific occasions: the invasions by the Syro-Ephraimite coalition and by Sennacherib. It is, rather, along two other lines that we must trace his message of hope. There is, first, the doctrine of the remnant, which, though not peculiar to his teaching, was an important feature of it. He gave one of his sons the symbolic name Shear-jashub, 'a remnant shall return' (or, 'repent'). It was, no doubt, as an acted prophecy that he took this son with him when he went to confront Ahaz with his appeal for faith (7^3).[1] But the remnant was not simply an element in Isaiah's teaching, but an outcome of it. When the rulers and the mass of the people were unresponsive to his prophecies, there were those few who did respond and become his disciples (8^{16-18}). In those who received Yahweh's word in faith Isaiah saw the cornerstone of the new Zion, the nucleus which was an Israel within Israel (28^{16}). The other element of hope is found in the promises of a coming king. For Isaiah, the coming age of bliss was one in which a well-ordered society would be ruled by a righteous king. The three most important passages are 7^{10-17}, 9^{2-7}, 11^{1-9}. The interpretation of the first of these raises complex problems; but it seems probable that, faced with a threat

---

[1] Similarly, a little later in the same crisis, Isaiah announced as the name of a child as yet unborn, Maher-shalal-hash-baz, 'speedy spoil, hasty prey', predicting the disasters which would befall Judah's enemies.

to the royal house (7^6), Isaiah predicts the birth of a prince as a confirmation of Yahweh's promises to the Davidic dynasty and of His presence with His people (Immanuel = God with us). The other two passages (which need not be regarded as later than Isaiah) give some indication of the character of a coming ideal king and of his rule. Their background is the ancient conception of the king as a channel of divine blessing; but they also point forward to the subsequent development of Messianic prophecy in Israel, which they profoundly influenced.

These four prophets, Amos, Hosea, Micah, and Isaiah are of historical importance because of the information which they provide about the conditions and events of their time. In the sphere of religion they stand out as witnesses to the ancient traditions and standards of the covenant community and also as men who saw in the crises of their time not sheer catastrophe but part of the continuing righteous purpose of Yahweh for His people.

From Manasseh to the Fall of Jerusalem (687–587 B.C.).

At the end of Hezekiah's reign Assyria was supreme in western Asia. Under Esar-haddon, the successor of Sennacherib, Egypt was successfully invaded in 671. Not long afterwards an Egyptian revolt was suppressed by Esar-haddon's son Ashurbanipal. The climax of his campaign was the capture and sacking of the ancient city of Thebes in 663 (cf. Nahum 3^{8-10}, where No-amon is Thebes). In such a situation Judah's subjection to Assyria could not be otherwise than complete. The religious consequence of this was the undoing of the reforms of Hezekiah, whose son and successor Manasseh (687–642) served the Deuteronomistic historian as an object lesson in apostasy (2 Kings 21^{1-18}). The high places and the fertility cult were restored; pagan altars were installed in the Temple; as in the reign of Ahaz (2 Kings 16^{17}), the Assyrian astral cult and human sacrifice were practised, as were various forms of divination and necromancy; and loyal adherents of the national religion were liquidated. The Chronicler states that on one occasion the Assyrians took Manasseh to Babylon in chains (2 Chron. 33[10 f.]). His narrative should not be too lightly written off as unhistorical: Manasseh may have been guilty of some act of disloyalty for which this was the punishment. But the subsequent account of a religious reform (2 Chron. 33[15 f.]) cannot easily be reconciled with other evidence of continuing apostasy.

Manasseh's son Amon (642–640) appears to have followed his father's policy (2 Kings 21[20-22]). When, as the result of a palace intrigue, he was assassinated, 'the people of the land'[1] promptly put the conspirators to death and secured the throne for Josiah, the 8-year-old son of Amon (2 Kings 21[23 f.]). What lay behind this internal conflict is uncertain. It has been surmised that the assassins were opposed to subservience to Assyria, or even in the pay of Egypt. However that may be, the new king (640–609) reigned during a period of drastic changes in the international scene, which enabled him to carry out a more independent policy than had been possible for his two predecessors.

Assyria's power was already on the wane. Within a decade of the fall of Thebes, Egyptian independence was successfully reasserted by Psammetichus I. Nearer home, Ashurbanipal had to suppress rebellion in Babylon and to overcome the Elamites and other trouble-makers on his borders. After his death (633?) the situation rapidly deteriorated. In 626 the Chaldaean Nabopolassar won freedom for Babylon and became the first ruler of the neo-Babylonian empire. In the east, the Medes made damaging attacks on Assyrian territory. The Scythians, barbarous tribesmen who had moved southwards from the region of the Caucasus, are said by Herodotus (*History* 1, 103–6) to have ravaged the western parts of the Assyrian empire as far as the borders of Egypt. In spite of the scepticism with which this statement has often been treated, it is possible that some at least of the references in the early chapters of

[1] The meaning of this phrase has been much discussed. In the pre-exilic period it probably denotes the free citizens of the country, by contrast with the royal family, the palace officials, the priests, and the prophets.

Jeremiah (e.g. 1^{14}, $4^{6\,f.,\,13\,ff.}$) to enemies from the north are allusions to the Scythians and that their onslaught provides the historical background of the prophecies of Zephaniah. In any event, the Scythians were one more unsettling factor in an already disturbed international situation. As Babylonian and Median pressure on Assyria increased, Psammetichus of Egypt came to the help of the tottering empire. Presumably his aim was to prevent a stronger power from dominating western Asia and threatening Egypt. But Assyria was doomed. In 614 the Medes captured Ashur; and in 612 Nineveh fell to a combined assault by the Babylonians and their allies. The Assyrians attempted to continue the fight with Haran as their centre of government; but Haran fell in 610. Necho II of Egypt continued his predecessor's policy of support for the Assyrians. But, even with substantial Egyptian help, an attempt in 609 to recapture Haran was quite unsuccessful. Assyrian power was now completely extinguished.[1]

All but the last act of this international drama was to the advantage of Josiah and his country. The long period of sub-servience to Assyria was over; and, even before the final collapse of the empire, it was possible for Judah to adopt an independent policy. This policy had three aspects: (a) political freedom from Assyrian control; (b) the opportunity, as a consequence of political freedom, to purify the cult from Assyrian influence; (c) the attempt to reunite with Judah the northern territory which had broken away from the house of David at the beginning of Rehoboam's reign, and which had been part of the Assyrian empire for two centuries.

The biblical narratives devote most of their attention to the second of these. Both 2 Kings 22–23 and 2 Chron. 34–35 describe important reforms in religious practice during Josiah's reign. The account in 2 Kings records the discovery, in the eighteenth year of the king's reign, of 'the book of the law' (22^8), the effect which the reading of this book had upon

---

[1] Valuable information about the events of this period is provided by the cuneiform record known as the Babylonian Chronicle. See *ANE*, pp. 202 f.; *DOTT*, pp. 76 f.

Josiah, and the religious reforms which he proceeded to carry out. But 2 Chronicles speaks first of a reform in the twelfth year of the king's reign (34^{3-7}), followed, in the eighteenth year, by the discovery of the book of the law and by further reforms. It may be that there were some early reforms, made possible by the decline of Assyrian power, and that this policy was intensified and developed when the book of the law came to light. At all events, the discovery of the book was a turning-point, not only in Josiah's reign, but in the religious history of Israel.

The reform, which was undertaken with the full weight of royal authority behind it, was a thoroughgoing programme of cultic purification and centralization. Pagan practices were abolished, including the astral cult and human sacrifice. The high places were suppressed; and thus the Jerusalem Temple was made the sole legitimate sanctuary, a measure which enhanced the prestige of the Zadokite priesthood there, and left the country Levites, now deprived of their previous status and means of livelihood, in a subordinate position (2 Kings 23^9). The reform was inaugurated by a renewal of the covenant (2 Kings 23^{1-3}; 2 Chron. 34^{29-32}). A national celebration of the Passover was also held, by royal command, at Jerusalem (2 Kings 23^{21-23}; 2 Chron. 35^{1-19}). The extension of the reform to the northern territory (2 Kings 23^{15}) was a religious measure with obvious and understandable political implications: the conditions which favoured the purification of the cult were also an encouragement to Josiah to attempt to restore the lost unity of the Davidic kingdom.

The measures adopted by the reformers correspond closely to some of the leading provisions in the Deuteronomic law code (Deut. 12–26), in which two of the dominant themes are the purification of the cult from alien features and the centralization of sacrificial worship at one legitimate sanctuary (12^{1-14}). Although that sanctuary is not named, it is undoubtedly Jerusalem. Deuteronomy directs that the Levites from the provincial shrines should be allowed to perform priestly functions at the central sanctuary and to receive the appropriate maintenance (Deut. 18⁶⁻⁸). The statement that they did not actually continue their priestly functions after Josiah's reform (2 Kings 23⁹) reflects a discrepancy between ideal and practice which does not disprove, but on the contrary points to, the fact that the Deuteronomic laws provided the basis for the reform.

The reform movement under Josiah was not by any means a new phenomenon. It had antecedents, of which the most noteworthy was Hezekiah's reform. But it did represent the formulation and application of ancient ideals and standards in a thoroughgoing fashion, in circumstances which were peculiarly favourable, and with the full weight of royal authority. Similarly, much of the material in Deuteronomy is ancient; but it is presented in a new idiom. There are passages which resemble some of the provisions of the Book of the Covenant (see above, p. 37), but with important differences (cf. Exod. 21² ff. and Deut. 15¹² ff.; Exod. 21¹² ff. and Deut. 19¹ ff.; Exod. 23¹⁰ f. and Deut. 15¹ ff.). The polemic against pagan practices was no new feature; but in Deuteronomy it includes particular reference to elements which had been prominent during the century preceding Josiah's reign: the astral cult and human sacrifice. Cultic centralization was also an old ideal presented in a new way. If it is true that in the early days after the settlement Israel was organized as an amphictyony (see above, p. 44), then the idea of a central sanctuary was not in itself a novelty; and there had been at least one earlier attempt, under Hezekiah, to suppress local shrines. But the effective limitation of sacrificial worship to the central sanctuary, as prescribed by Deuteronomy and carried out by Josiah, was a new and far-reaching measure. As we have seen, centralization had important consequences for the Levitical priesthood. For the ordinary Israelite, one of its chief results affected the killing of domestic animals for food. Previously this had been permissible only as a sacrificial act; but now that there was only one centre for sacrifice, provision was made for the non-sacrificial slaughter of animals at places other than the central sanctuary (Deut. 12¹⁵ f., 20⁻²⁵). Yet another effect of the law of centralization was the celebration of the Passover, not in the home, but at the central sanctuary (Deut. 16[1-8]; 2 Kings 23[21-23]).

Along with the blending of old and new there probably went a blending of northern and southern elements. Although the reform secured the position of the Jerusalem sanctuary and the dominance of its Zadokite priesthood, there is much to be said for the view that Deuteronomy incorporated material from the traditions of northern Israel. The parallels to the Book of the Covenant, some of which have been cited above, are among the most striking of a number of resemblances to the northern (Elohistic) material in Genesis–Numbers. There are also points of contact with the northern prophet Hosea; and the attitude adopted to the monarchy has been thought to reflect the attitude of northern reformers. It is, indeed, plausible to suppose that when the northern kingdom fell, refugees found their way to Judah, and that by this means northern traditions and legal formulations came to be incorporated with those of Judah. At all events, it is an undivided Israel that Deuteronomy presupposes: the purified and centralized cult is the expression of the worship of all Israel, the people whom Yahweh chose 'to be a people for his own possession' (Deut. 7[6]). As we have seen, it appears to have been Josiah's aim, in the carrying out of the reform, to include in its scope, and therefore under his own authority, the territory north of Judah.

The exhortations and regulations for the maintenance of a pure and centralized cult and for the ordering of Israel's life are prefaced in Deuteronomy by two retrospective passages (1–4 and 5–11), which recall and interpret Yahweh's deliverance and guidance of Israel in the Mosaic age. These provide a theological basis for the laws. (a) Yahweh is One: 'Hear, O Israel: The LORD our God is one LORD' (6[4]). The sole lordship of Yahweh and the unity of His being could be obscured by the worship of other deities and by variations and corruptions in the worship of Yahweh at local shrines (cf. Jer. 2[28]; 'for as many as your cities are your gods, O Judah'); hence the law of centralization. (b) Yahweh has *chosen* Israel as His people, not because of their numerical importance, but because of His love (7[7 f.]). This expression of the relationship between Yahweh and Israel in terms of *election* is of great importance in Deuteronomic teaching.[1] It rules out any thought that the people of Israel have deserved or earned Yahweh's gifts; and it emphasizes what should be the motive of their response to Him. Gratitude for Yahweh's undeserved goodness should move them to obedience to his commandments. They should respond in whole-hearted love to Him (6[5]); and as He had compassion on them in their affliction in Egypt they must show a like compassion to the afflicted, the needy, and the defenceless (10[18 f.]). (*c*) The land is theirs, not of right, but because Yahweh has given it to them as an inheritance; and they hold it only on condition that they remain faithful and obedient (8[7-20]).

Although many of the immediate effects of the reform were undone after Josiah's death, its influence continued in important ways. It meant much for the future of Israel's religion that there should have been this reassertion, in the highest circles in the land, of the ancient standards of Yahwism, only a generation before the fall of Jerusalem. Among the exiles and later in the returned community, the teaching of Deuteronomy survived. Moreover, the influence of Deuteronomy is evident in many other parts of the Old Testament. This is true, above all, of the historical narrative of Joshua–Judges–Samuel–Kings. At the beginning of that narrative, Israel received the land as its inheritance, and at the end lost it because of apostasy and disobedience. National prosperity is interpreted as the reward of faithfulness, and national affliction as the punishment of infidelity to Yahweh; kings are praised for adherence to Deuteronomic standards, and blamed if they departed from them. Deuteronomy, which thus provided standards by which Israel's history was interpreted, became a foundation document of the Judaism which was to survive the Exile.

---

[1] Parallel to it is the thought of the election of Jerusalem, 'the place which the LORD your God will choose to make his name dwell there' (Deut. 12[11], etc.). We may also compare the thought of the choice of David and his dynasty (Ps. 89[3, 19 f.]).

Judah's brief period of political freedom ended in 609, when Josiah was killed by the Egyptians at Megiddo. Necho II was on the way to Carchemish to help the Assyrians against their enemies. Josiah had clearly renounced Assyrian overlordship: and it is not surprising that he should have attempted to obstruct the Egyptian advance; but the attempt ended in disaster for him and for his country.[1] Josiah's son Jehoahaz was made king by 'the people of the land'; but three months later his reign came to an abrupt end when Necho deposed him and deported him to Egypt. Eliakim, an older son of Josiah, was now installed as king by Necho and given the regnal name of Jehoiakim (2 Kings 23$^{33\,f.}$; cf. Jer. 22$^{10-12}$, where 'Shallum' refers to Jehoahaz).

For the time being, Egypt dominated the country. The national resources were depleted by the payment of heavy tribute and by Jehoiakim's extravagant building enterprises. Solomon's faults seem to have been reproduced in a king who, with no claim to Solomon's wisdom, had to conduct national policy in a dangerous and rapidly changing situation. Jeremiah has left a memorable portrait of him in a trenchant denunciation which points the contrast with Josiah (Jer. 22$^{13-19}$). But there must have been many in Judah who were keenly aware of another contrast; that between the rewards of obedience promised in Deuteronomy and the violent death of the reforming king. Since Josiah's reign had ended in disaster while he was still a relatively young man, did it not follow that his religious policy had been misguided? It is understandable that there was a revival of pagan practices (e.g. Jer. 7$^{16-18}$).

Jehoiakim was soon deprived of the support of his Egyptian overlord. In 605 the Babylonian crown prince, Nebuchadrezzar,

[1] See 2 Kings 23$^{29\,f.}$; 2 Chron. 35$^{20-24}$. In Kings it is stated that Necho 'went up to [R.V. "against"] the king of Assyria'. The Babylonian Chronicle makes it clear that he went as Assyria's ally. The Chronicler describes how Josiah was mortally wounded in battle. Since Kings says nothing of this, some have inferred that Necho summoned Josiah to Megiddo and punished him as a contumacious vassal of Assyria. But there is no need to discount the Chronicler's version.

decisively defeated the Egyptians at Carchemish and again at Hamath, and drove them back through Syria and Palestine (2 Kings 24⁷). The death of Nabopolassar made it necessary for Nebuchadrezzar to leave the scene of operations and return speedily to Babylonia to ensure his own succession. But Babylonian pressure in the west was not seriously interrupted. For the second time in less than five years there was panic in Jerusalem. Jehoiakim had perforce to become a vassal of Babylonia. But some three years later, encouraged perhaps by the failure of a Babylonian attempt to invade Egypt, he rebelled. Nebuchadrezzar was unable to act decisively at once. When, in 598, the Babylonians invaded Judah and besieged Jerusalem, Jehoiakim was already dead. His son and successor, Jehoiachin, had to capitulate in 597 after a reign of only three months. His capital was robbed of its treasures; and Jehoiachin was deported together with many leading members of the court and of the upper classes (2 Kings 24⁸⁻¹⁷).

The crises of the previous decade, culminating in the fall of Jerusalem, had reduced the strength and probably the territory of Judah. But the hardest blow which the Babylonians struck at her was to replace Jehoiachin by his uncle, Mattaniah who received the regnal name of Zedekiah. Lacking both resolution and judgement, he was not the man to guide the country's policy in the years which lay ahead. Egypt revived her old practice of fomenting trouble in western Asia and of actively intervening there. Within Judah there were two contending parties: those who favoured continued submission to Babylon; and those who, presumably under Egyptian influence, advocated revolt. Among the exiles, too, there were those who hoped and schemed for the downfall of Babylonia (Jer. 28 ¹⁻¹¹). Left to himself, Zedekiah would not have been a troublesome vassal; but he was unable to withstand the pressure of the anti-Babylonian party. In 589 he revolted. The following year his land was invaded and his capital besieged (2 Kings 25¹). Of the other strongholds, only Azekah and Lachish were able to maintain some sort of resistance (Jer. 34⁷). The military situation is illustrated in a vivid but fragmentary fashion by some of the Lachish Ostraca.[1] Without help from outside, effective resistance could not be prolonged indefinitely. There was, indeed, a brief interruption of the siege when an Egyptian army marched north; but the Babylonians quickly repelled the intruders (Jer. 37[5-10]), and Jerusalem was again invested. In the summer of 587 the wall was breached and the city fell. Zedekiah tried to escape, but was overtaken at Jericho

Ostrakon from Lachish.

and taken to Nebuchadrezzar's base at Riblah. After seeing his sons put to death, he was blinded and deported to Babylon. Shortly afterwards, Nebuchadrezzar sent to Jerusalem Nebuzaradan, the commander of his guard, under whose orders the Temple and city were plundered and burned, the walls broken down, and, of those members of the upper classes who had escaped the deportation of 597, many were carried off to

[1] The Lachish Ostraca are twenty-one inscribed potsherds discovered in 1935 and 1938 on the site of the fortress town of Lachish (Tell ed-Duweir). They include letters sent to the military governor of Lachish from one or more of the Judaean outposts. In one of the letters it is reported that the fire signals from Azekah can no longer be seen; presumably the stronghold had by then fallen. See *ANE*, pp. 212-14; *DOTT*, pp. 212-17.

Babylon (2 Kings 25$^{1-21}$). Thus the royal line of David was deposed and humiliated; the capital and its Temple, which Yahweh had chosen to cause His name to dwell there, were despoiled and desecrated; and considerable numbers of the people, including most of the leaders, removed from the land which Yahweh had given to their forefathers.

Gedaliah, a Judaean noble, was installed by the Babylonians as governor, with Mizpah as his administrative centre; but his régime did not last long. He was treacherously assassinated by Ishmael, a member of the royal house, who was incited by Baalis, king of Ammon. Gedaliah's associates were afraid that the Babylonian retribution for this act would fall on them. They therefore resolved to decamp to Egypt; but before doing so they consulted the prophet Jeremiah, who had been exempted from deportation because of his pro-Babylonian attitude. He advised against flight; but his appeals were rejected, and he was himself carried off to Egypt against his will (2 Kings 25$^{22-26}$; Jer. 40–44). The further deportation of Jews to Babylonia which took place in 582 (Jer. 52^{30}) may have been Nebuchadrezzar's reaction to the murder of his representative.

Describing the horrors and humiliation of this period, a Hebrew poet wrote, 'Her [Zion's] prophets obtain no vision from the LORD' (Lam. 2^9); and, in a psalm which is often thought to refer to the same catastrophe, another wrote, 'There is no longer any prophet' (Ps. 74^9). The military and political reverses must indeed have had a devastating effect on religious life in Judah.[1] But the previous half-century had been a time of intense prophetic activity. The Deuteronomic legislation contains a strong assertion of the importance of prophecy in the national life and of the distinction between the prophets of Yahweh and pagan diviners, necromancers, and the like

[1] It is, however, noteworthy that, when Gedaliah was assassinated, Ishmael's victims included seventy out of a party of eighty men who were on their way from Shechem, Shiloh, and Samaria 'to the temple of the LORD', i.e. presumably to the ruined Temple at Jerusalem (Jer. 41$^{4-9}$). When the Babylonians had done their worst, the sanctuary in Zion could still draw worshippers, even from the great northern centres.

(Deut. 18⁹⁻²²). On the other hand, the varied character of prophecy within Israel is indicated by the Deuteronomist's warning against prophets who may tempt Israel to apostasy (Deut. 13¹⁻⁵). Although these passages reflect experience of the prophetic movement over many generations, they are also pointedly relevant to the closing decades of the independent kingdom of Judah. The book of Jeremiah provides ample evidence of the varied prophetic activity of that time. Jeremiah uttered devastating denunciations of prophets who, for all their volubility, had no authentic word from Yahweh (Jer. 23¹⁵ ᶠᶠ·). Both in Jerusalem and among those who were deported to Babylonia in 597 there were nationalistic prophets who fomented resistance and encouraged the people of Judah and their leaders to believe that the fall of the great empire was imminent (Jer. 27¹⁴ ᶠᶠ·, 28, 29⁸ ᶠᶠ·, ²¹ ᶠᶠ·). The contrast between such men and Jeremiah himself is pointedly reminiscent of the encounter between Micaiah the son of Imlah and the prophets who encouraged Ahab to attempt to recapture Ramoth-gilead (1 Kings 22); it is, indeed, a contrast which recurs throughout the history of Hebrew prophecy before the Exile.

There are five prophetic books which are related, in whole or in part, to the period under consideration: Jeremiah, Ezekiel, Zephaniah, Nahum, and Habakkuk. Ezekiel is the great prophet of transition. Part of his ministry was exercised before the final fall of Jerusalem; but since he lived and worked among the exiles, and since in his work and teaching he adumbrated important future developments, it is best to consider him in the context of the Exilic period.

The activity of Zephaniah belongs to the reign of Josiah, and probably to the earlier part of it. It has often been assumed that the Scythian onslaughts (see above, pp. 123 f.) were the occasion of his prophecies; but, though this is a possibility, it must be admitted that Zephaniah makes no direct and explicit reference to the Scythians. But that is a matter of subordinate importance. The three dominating themes in Zephaniah's teaching are these: first, denunciation of disloyalty to Yahweh in Judah and Jerusalem, as expressed both in pagan cultic practices and also in irreligious self-confidence and wrong-doing (1^{4-6}, $8^{f.}$, 12, 3^{1-4}); second, the prediction of divine retribution when 'the great day of the LORD' comes ($1^{2 f.}$, 7, $14-18$, $2^{1 f.}$); third, the assurance that a remnant, consisting of the meek and righteous, will survive the catastrophe (2^3, $3^{12 f.}$). The first of these echoes much that had been said by earlier prophets. If the oracles are rightly dated in the decade 630–620, this was part of the prophetic reaction against the apostasy of Manasseh's reign, and helped to prepare the way for Josiah's reform. The second is a development of an idea which was already present in the teaching of Amos and Isaiah (Amos 5^{18-20}; Isa. 2^{6-22}; see above, pp. 115, 120 f.) and which was to feature largely in later predictions of future judgement and deliverance. The third is also connected with an element in Isaiah's ministry (see above, p. 121) which is found in varying forms elsewhere in the Old Testament: the thought of a nucleus in Israel which is the true Israel and for which Yahweh will still have a purpose when the expected catastrophe is past.

The book of Nahum takes us on to the pivotal event of the next decade: the fall of Nineveh. For the most part it is an exultant description of the overthrow of 'the bloody city', probably composed in anticipation of the event, though some scholars think that it is not a prediction but a celebration of what had already taken place. Liturgical characteristics have been discerned in the book; and it has been inferred that Nahum was a cultic prophet, hurling ritual curses against the hated enemy. Because his oracles contain no word of judgement on Judah, he has usually been described as a nationalistic prophet, and even compared to the prophetic adversaries of Jeremiah. But this is an unjustified inference. In Nahum's savagely brilliant poetry we may sense the jubilant relief with which men hailed the fall of a ruthless empire. Further, in presenting Yahweh as the ultimate author of retribution on Assyrian tyranny, Nahum is at one with the classical prophetic interpretation of history.

The interpretation of the message of Habakkuk is beset by problems to which a variety of solutions has been offered. The central question in the book is, 'Why does Yahweh permit wrongdoers and violent men to oppress the righteous?' This is first put in 1^{2-4}, where the reference is probably to ruthless conduct by Jews rather than to the actions of a foreign power (Egypt or Assyria). The situation presupposed is probably the early part of Jehoiakim's reign. Yahweh's answer to the complaint is that He is raising up the Chaldeans (i.e. the neo-Babylonian empire) to punish the wicked (1^{5-11}). A further complaint about the persecution of the righteous by the wicked (here probably a foreign power, i.e. the Chaldeans) is answered by the assurance that the divine purpose will be made plain, and that the righteous man will live by his faithfulness (2^{1-4}). The five 'woes' which follow are very much in the classical prophetic style (cf. Isa. 5^{8-24}). The closing poem (3) is a vivid description of Yahweh's coming in judgement, combined with a sublime expression of trust in His faithfulness. It is usually thought that this poem was not originally connected with the prophecy, but was added to it from some collection of psalms.[1] Of Habakkuk himself we have no direct knowledge; but it has been argued that liturgical features in the prophecy indicate that he was a cultic prophet, and that chapter 3 is the climax of a prophetic liturgy. However that may be, in grappling with the problem of reconciling the sovereign righteousness of Yahweh with the outward course of events, Habakkuk presents a theme which appears in varying forms elsewhere in the Old Testament (e.g. in the book of Job, and in many of the laments in the Psalter), but not least in the thought and experience of his own great contemporary, Jeremiah.

Jeremiah came of a priestly family in the village of Anathoth, near Jerusalem, the home of the priest Abiathar, who was dismissed by Solomon. He became aware of his vocation to be a prophet in 627/6 B.C. (Jer. 1^2); and his ministry spanned the last forty years of Judah's independent existence and continued (for how long we do not know) into the bleak period that followed. Decisive events in the history of the time mark off four

[1] The Commentary on Habakkuk, which is one of the most notable of the Dead Sea Scrolls, contains no allusion to chapter 3.

main phases, which were different in character from each other.

(*a*) The first of these lasted from the time of Jeremiah's call until Josiah's reform. It is represented by the bulk of the material in Jer. 1–6. Jeremiah inveighed against religious corruption and apostasy in Judah, and predicted that an enemy from the north would be Yahweh's instrument to punish His people. The tone is set by the double vision of the almond branch and the boiling pot (1^{11-16}). These are both tokens of imminent catastrophe, for the almond branch is a sign that Yahweh will fulfil His word of doom. The enemy from the north ($1^{14 f.}$, 4^6, 6^1) has often been equated with the marauding Scythian hordes (see above, pp. 123 f.).

(*b*) From the reform until the death of Josiah is something of a mystery period in Jeremiah's prophetic ministry. It is difficult to assign to it either oracles or narratives; and there has been much disagreement about the attitude which Jeremiah may have adopted to Josiah's religious measures. On the one hand, Jer. $11^{1 ff.}$ reads like a call to Jeremiah to promote the reform. Further, the abuses against which he had protested were those which the reform sought to suppress. But, on the other hand, the book of the law emphasized the importance of right sacrificial practice, whereas Jeremiah was the most outspoken of all the prophets in his condemnation of sacrifice (7^{21-26}); the book exalted the importance of the Temple, as the place where God had caused His name to dwell, and as the one legitimate place for sacrifice, whereas Jeremiah denounced false trust in the Temple (7^{1-15}, 26); the reform involved the suppression of the local shrines and ensured the dominance of the Zadokite priests of Jerusalem, but Jeremiah came from a priestly family at a local shrine associated with Abiathar, whom Zadok superseded. It is widely held that Jeremiah was at first a supporter of the reform, but later came to feel that it was a failure, because it was concerned only with outward conformity. If this were so, one could wish that the prophet had had time or opportunity to ponder Deut. $6^{4 f.}$ But, in fact, the unqualified approbation which Jeremiah gave to Josiah after Josiah's death (Jer. 22$^{15f.}$) does not suggest that he had come to disapprove of the policy of reform.

(c) The period of rather more than twenty years from the death of Josiah to the final fall of Jerusalem and the second deportation was for Jeremiah a time of conflict, accusation, and persecution. This is recorded in much of the material contained in 7–39, consisting partly of prophetic oracles, partly of Jeremiah's account of his own inner struggles, and partly of narratives recorded by his friend and amanuensis, Baruch. As in most other prophetic compilations, strict chronological order is not followed; but there is such a wealth of biographical narrative, that it is not difficult to reconstruct the general progress of events. At the beginning of Jehoiakim's reign, when the nation must have been overwhelmed by the double shock of the death of Josiah and the deportation of Jehoahaz, and, perhaps, disposed to look to the Temple as the pledge of security, Jeremiah intervened with an abrupt assertion that, Temple or no Temple, their immorality was an offence to Yahweh, and that Jerusalem and its sanctuary would be destroyed, like Shiloh five centuries earlier (7$^{1-15}$; 26). After the battle of Carchemish, when the international situation had been transformed and Jehoiakim's foreign policy was in ruins, Jeremiah dictated to Baruch the oracles which he had spoken in previous years. Baruch then read them publicly in the Temple precincts. The hope was that at this time of crisis the people would be moved to repentance. In the event, the king destroyed the manuscript; and Baruch and Jeremiah narrowly escaped punishment. It is evident from the details of this narrative (36) that Jeremiah had influential friends at court. But for their help, he would no doubt have been put to death, either then, or during the ten uneasy years of Zedekiah's reign. Although Zedekiah on occasion consulted him, his advocacy of submission to the Babylonians brought him under suspicion as a traitor. He was arrested, beaten, and imprisoned (37^{11}–38^{13}). When the city finally fell, the Babylonians released him.

(d) Chapters 40–45 cover the fourth period of Jeremiah's ministry, which began during Gedaliah's brief governorship at

Mizpah, and ended, presumably, in Egypt, where he continued to protest against the pagan practices of the Jews who had settled there.

In Jeremiah's teaching we find, as in that of his eighth-century predecessors, the assertion that Yahweh has acted and is acting in history, the appeal to the ancient bond between Yahweh and His people, and the denunciation of corrupt worship and social injustice. Both in some of his sayings (e.g. 2^2, 3^{22}) and in the note of personal suffering which runs through so much of his prophetic ministry, there is a close affinity with Hosea. There is, too, alongside his predictions of seemingly inexorable doom, more than one element of promise in his teaching. As the potter can remake a marred vessel, so Yahweh is ready to remake the nation if it turns from evil ($18^{1\,ff.}$). Chapters 30–31 (sometimes called 'The Little Book of Comfort') contain messages of hope, and, in particular, the expectation that the exiles of the northern kingdom will be restored and Israel and Judah be reunited. After the first deportation, Jeremiah saw in the exiled community the pledge of the continuation of Yahweh's purpose for His people (24, 29). But it is in the prophecy of the New Covenant ($31^{31\,ff.}$) that Jeremiah's most characteristic expression of hope is found. This is not a radically new and individualistic concep-tion of the covenant: the New Covenant is still 'with the house of Israel and the house of Judah'; but within its communal framework there is the personal, individual, and inward knowledge of Yahweh and of His will; and its foundation is Yahweh's free forgiveness of his people's sin.

Jeremiah is often described as the great 'individualist' of the Old Testament. But his 'individualism' is not some abstract principle which he felt called to enunciate. It is rather the expression, on the one hand, of his sense that he had been predestined by Yahweh for his prophetic task (1^{4-10}), and, on the other, of the isolation which he experienced when his pre-dictions of doom were disbelieved and for years remained un-fulfilled, when the very people for whose repentance he longed treated him as a laughing-stock or a traitor. This experience of failure and isolation drove him back upon Yahweh, sometimes in bitter protest. In the passages known as the Confessions of Jeremiah (11^{18}–12^6, $15^{10\,ff.}$, $17^{9\,f., 14-18}$, 18^{18-23}, 20^{7-18}) we have glimpses of this aspect of Jeremiah's ministry. It is, indeed, part of his prophetic task. If he had been an out-and-out individualist, he could have abandoned the task of prophesying to his people, and his sacrificial struggle would have been at an end. But this he could not do. Jeremiah comes at the end of an age; but he points forward to a new age in which there would be a new understanding of the common life of the people of Yahweh and of the place of the individual in it.

VI

THE EXILE AND AFTER

THE sixth century B.C. was a momentous turning-point in the history of the Near East. The neo-Babylonian empire, which had been established by the aggressive policy of Nabopolassar and his son Nebuchadrezzar, was the last great Semitic power in biblical times. When it was overthrown by the Persians in 539, there began a period of more than a millennium during which non-Semitic powers dominated Asia. The Persian empire lasted for two centuries. It was ended by the victories of the Macedonian conqueror, Alexander the Great, whose aim it was to disseminate Greek culture in the territories which he annexed. Although his empire fell apart soon after his death (323 B.C.), the process of Hellenization continued in western Asia during the centuries which followed, and on into the period when Roman rule was established in Asia Minor, Syria, Palestine, and Egypt. It was not until the rise of Islam in the seventh century of the Christian era that a Semitic power was again in control.

The sweeping changes which took place during the closing centuries of the Old Testament period involved not only the obliteration of territorial national boundaries, but the fusing of cultural and religious traditions. It is, therefore, all the more remarkable that the fall of Jerusalem and the Exile did not mean the end of Israel, but that there was a recognizable continuity of religious tradition and of community life. As the changes which were consequent upon the settlement in Canaan had brought danger, challenge, conflict, and enrichment, so the more painful changes which were involved in the loss of national independence, and the long process of adjustment to political impotence and exposure to powerful alien influences, brought new hazards and new gains.

To trace the history of the Jews[1] during this period is a baffling task. Historical sources are scanty; and those which do exist present formidable problems to the historian. But, although we cannot always reconstruct the precise sequence of events, certain broad lines of development are plain.

The Exile and its Consequences

The Exile was not in any strict sense a 'captivity'. It is clear from the book of Ezekiel (8[1], 20[1]), that the exiles could meet and confer freely; and the implication in Jeremiah's letter to those who had been deported in 597, that they would be able to carry on their normal occupations, was undoubtedly realized in actual experience. It is also important to remember that the Exile did not involve the removal of anything like the entire nation. Even after three deportations, those who had been settled in Babylonia must have been only a fraction of the total population of the kingdom of Judah.[2] The importance of the Jewish community in Babylonia lay not so much in its size as in its quality. It included many (perhaps most) of the leaders in various departments of life. Among them were fostered the supremely formative influences on later Judaism. It was when

[1] Strictly, the word 'Jews' ought to denote members or descendants of the tribe of Judah. It does not seem to be used in the Old Testament to distinguish those who belonged to the kingdom of Judah from their neighbours in the northern kingdom, but is first used after the northern kingdom had fallen (2 Kings 25[25]; Jer. 38[19], 52[28-30]). From this time onwards it is used of those who shared the national and religious inheritance of Israel, in contrast to 'Gentiles', but not necessarily only of those who claimed descent from Judah. St. Paul, who was 'of the tribe of Benjamin' (Phil. 3[5]), called himself 'a Jew' (Gal. 2[15]).

[2] According to Jeremiah 52[28-30], Nebuchadrezzar deported 3,023 'Jews' in 597, 832 'persons' in 587, and 745 'persons' in 582: a total of 4,600. 2 Kings 24[14] (which is generally regarded as a late insertion) speaks of 10,000 'captives' in 597; and 2 Kings 24[16] (referring to the same deportation) records the removal of 8,000 'men'. Even if we accept the higher numbers given in Kings, and assume that they refer only to males capable of bearing arms, the total must have been considerably less than the population of Judah at the time, which has been variously estimated at 125,000 and 250,000.

parties from this community returned to Jerusalem after the fall of Babylon that life in the homeland began to be re-organized.

The plight of Jerusalem and Judah and of those who re-mained behind immediately after the catastrophe is movingly described in the book of Lamentations. For the rest of the exilic period, we have no direct and certain evidence. It seems likely that after the assassination of Gedaliah the territory was governed from Samaria, the administrative centre for the northern territory. There is no evidence that Nebuchadrezzar brought into the country settlers from other parts of his em-pire, as the Assyrians had done in the north after the fall of Samaria; but there are some indications that Judah's neigh-bours, and particularly Edom (cf. Ps. 137[7]; Ezek. 25[12-14]; Obad. [10 ff.]) took advantage of Judah's helplessness. Living conditions for those Jews who remained in the land must have been miserable; and it is probable that many emigrated, hoping for better things. Those who carried off the unwilling Jeremiah settled at Tahpanhes, just beyond the Egyptian border (Jer. 43[7]). There were other settlements of Jews in Egypt (Jer. 44[1]),[1] which in later times was the home of in-fluential Jewish communities. Indeed, to speak of the exilic and post-exilic periods is misleading if it gives the impression that the Jews went into exile in Babylonia for half a century, after which they, or their children, returned. Some, indeed, did return; but many remained in Babylonia; and many others emigrated from Judah to other lands. The exilic age is the beginning of the age of the Dispersion, or Diaspora. From this point onwards we must reckon with the existence of Jewish communities scattered throughout Mesopotamia and countries surrounding the Mediterranean. Many of them looked with longing to the land of Israel, now ruled by alien powers, and

---

[1] The Elephantine papyri provide valuable information about the life of a military colony of Jews on the island of Elephantine, opposite what is now Assuan. This settlement dated from before 525 B.C. but unfortunately we do not know exactly when or how it was founded. See *ANE*, pp. 278–82; *DOTT*, pp. 256–69.

to the city of David and its once glorious sanctuary; but the ancient centre of Israel's life and worship remained for centuries under alien domination. Whether he lived in the homeland or in the Dispersion, the pious Jew had to practise his religion in conditions of political subjugation. For a religion which had been so closely linked to the political structure of national life, this was bound to have far-reaching consequences.

It has often been said that Israel went into exile a nation and returned a church. The saying is misleading, for Israel never lost her national consciousness, even if for long periods she was deprived of the principal means of expressing it. But it nevertheless points to an important truth; for clearly, if the unity and distinctiveness of the now scattered Jewish community were to be preserved and made evident, there had to be an emphasis on those religious usages which both demonstrated and strengthened the spiritual cohesion of Judaism and marked it off from its alien environment. Further, since so many Jews now lived outside Palestine, those religious usages which were independent of worship at the Temple acquired a new importance. There was an increased emphasis on the observance of the Sabbath, on circumcision, and on adherence to the dietary laws. There was growing concern for the formulation and application of the details of religious law, both so that the life of the religious community might be well ordered, and also so that the individual Jew might know what God required of him. There was also, of course, a passionate desire for the restoration of the Temple, and when that became possible, for the maintenance of worship there.

The Transformation of Prophecy

The fact that Israel's faith survived the national catastrophe was in part due to the fact that there had been prophets in Israel who predicted doom as Yahweh's punishment and discipline of His people. What might have been regarded as a sign that Yahweh had been defeated by the Babylonian deities was regarded as the triumph of His purpose. The prophets of doom, whose message had been so often rejected, were vindicated by events. But prophecy itself passed through important changes in the period which followed the fall of Jerusalem. As a movement, it lost its vitality and even sank into disrepute. This may be illustrated by two contrasted passages. In Zech. 13^{2-6}, the prophet is represented as a charlatan and a corrupting influence, who will be eliminated from the national life in the ideal future. In 1 Macc. 4^{46} (cf. 9^{27}, 14^{41}) it is implied that the absence of the gift of prophecy is a lack which will be made good in the ideal future. On the other hand, the prophetic teaching which has been preserved from the period of the Exile and later reflects important changes. We have already noted that the element of hope was by no means absent from the teaching of the pre-exilic prophets; but now that Israel's present experience was so depressing, that element was developed, and there came a new emphasis on the glory of the new order which God would inaugurate. Israel's need of penitence, discipline, and purification was not forgotten; but when Israel was under the power of tyrannous pagan rulers, and when it was often the pious in Israel who suffered most, it was natural that visions of the future should include the punishment of Israel's oppressors. Pre-exilic prophets had condemned the cultic practices of their contemporaries, but in an age when the right observance of the cult was one of the marks of the continuity and distinctiveness of Israel's life, neglect and slovenliness in the cultic practice could be a target of prophetic invective. In the past, the prophet had been primarily a man of the spoken word: the committing of his words to writing was secondary (Isa. $8^{1, 16}$; Jer. 36). But now there was an increasing tendency for the prophet to use the written word in the first instance to communicate his message. It may be that in the conditions of the time speaking in public was both more dangerous and less effective as a means of communication. This development in prophecy of a more consciously literary character, together with the elaboration of detailed predictions of the future, often in highly symbolic terms, led on to the type of literature known as apocalyptic, which though different from prophecy, was in some sense its heir.

Ezekiel

The great herald of things to come was Ezekiel, not only in the sense that he predicted the future, but also because, in the manner and content of his prophetic ministry, he foreshadowed many of the important religious developments which were characteristic of the age after the Exile. He, rather than Ezra, was the founder of Judaism. He not only pointed forward; but as we shall see, he represented some of the great elements in Israel's religious past.

The book which bears his name is outwardly impressive in its orderliness and symmetry and in the careful chronological arrangement of its contents. It purports to present the record of prophecies uttered in the Babylonian Exile between 593 and 571 B.C.; and for long this was not seriously questioned. Even when other prophetic books had been dissected and assigned to sundry authors and editors, this book continued to be regarded by most scholars as having come in its entirety from Ezekiel. Then came a period in which many extreme theories were advanced, assigning much of it to other hands, or presupposing complicated processes of editorial revision, or dating the book to a period much later than the Babylonian Exile, or maintaining that Ezekiel's ministry was not exercised in Babylonia but in Palestine, or at least was begun there. Such theories have been subjected to damaging criticism, and are now somewhat discredited. The account of Ezekiel's ministry and teaching given below is based on the view that he lived and worked among the exiles in Babylonia, at the period indicated, and that the bulk of the material in the book comes from him, though, like other prophetic collections, it owes much in its compilation, arrangement, and transmission to prophetic disciples.

Ezekiel was a priest who had been carried away in the deportation of 597, and who received his call to be a prophet in 593. His ministry fell into two unequal parts. From 593 until 587, in face of the unfounded optimism of the exiles, he predicted the fall of Jerusalem as inevitable, and denounced the corruption of the national life and worship. After the cata-strophe, in face of the despondency and fatalism of his hearers, he foretold the miracle of renewal and restoration. Thus he was prophet of both judgement and promise. In him we see exemplified the classical tradition of pre-exilic prophecy, but also another side of the prophetic movement which is less obvious in men such as Amos, Isaiah, and Jeremiah. He describes not only several extraordinary symbolic actions (e.g. 4, 5$^{1-4}$) but also a number of abnormal psycho-physical experiences such as temporary dumbness (3^{26}), paralysis (4), and the sensation of being transported bodily from Babylonia to Jerusalem (8^3). Exactly how such descriptions should be interpreted is uncertain; but it seems that Ezekiel had in-herited a larger share of the ecstatic side of prophecy than some of his predecessors. He gathered up in himself the fullness of the prophetic tradition and combined it with the priestly inheritance which was his by descent, and which appears in his concern about the minutiae of arrangements for worship. The chapters in which this is expressed (40–48) are an in-teresting blend of detailed description of an ideal future with that interest in laws and regulations which was a characteristic of the period immediately following. Further, in his predictions of the future Ezekiel not only sounded the note of confident hope which was heard again in Deutero-Isaiah and later prophets, but he made use of the grandiose and at times fan-tastic symbolism which was later developed in apocalyptic literature. He was a brilliant literary artist, and, as such, a precursor of the more literary phase into which prophecy was moving. There is a particular appropriateness in the fact that in his experience of the call to be a prophet he was commanded to eat a *roll of manuscript* (2^8–3^3).

The description of the prophet's call in 1–3 contains three of the central elements in his subsequent ministry.

First there is the vision, which the prophet had by the river Chebar, of a great storm-cloud from the north out of which appeared four living creatures bearing a throne-chariot (here already we may note apocalyptic characteristics) upon which was 'the appearance of the likeness of the glory of the LORD' (1^{28}). An awareness of the ineffable majesty of Yahweh is combined with a sense of His active presence. A similar vision is recorded later when Yahweh departs from the doomed city of Jerusalem (10). Throughout the prophecies, in a variety of ways, Ezekiel conveys this sense of the transcendent, holy God. Yahweh's presence is made known by the dazzling brightness which is His glory. He acts 'for the sake of His name' ($20^{9,\ 14,\ 22,\ 44}$), i.e. in order to demonstrate His true nature. Ezekiel is frequently addressed as 'son of man' (i.e. mere man, mortal man), a phrase which emphasizes the distinction between God and man.

Second, there is the emphasis on the individual. Ezekiel is called to perform the function of a watchman who warns of impending danger (3^{16-21}; cf. 33^{1-9}); but his responsibility is not simply for the community in general, but for the individuals of whom it is composed. Another aspect of this emphasis appears in 18. The despondent exiles say that they are doomed because of the sins of an earlier generation: 'The fathers have eaten sour grapes, and the children's teeth are set on edge.' Ezekiel denies this in a detailed argument which is more than an assertion of individual moral responsibility. The heart of the matter is that a new start is possible now, because Yahweh has no pleasure in the death of the wicked, but desires that he may 'turn from his way and live' (18^{23}).

This note of compassion needs to be borne in mind when we turn to the third element in Ezekiel's call and ministry; his mission to a rebellious people. He is warned not to expect any ready response (3^{4-11}). Their history is described as one long record of perversity and disobedience (20). Ezekiel paints a black picture of contemporary corruption (8). But beyond the catastrophic judgement lies renewal. Two memorable pictures stand out. One is of the active compassion of Yahweh, the divine Shepherd who goes out to seek and save His sheep (34). The other is of the bleached bones of the dead nation, brought together into bodies of flesh and sinew by the word of prophecy and made alive by the spirit of Yahweh (37^{1-14}). For Ezekiel, restoration was sheer miracle: not only the miracle of over-throwing other nations and thus preparing the way for the re-establishment of Israel's national life; but the miracle of giving Israel a new heart (11^{19}). This is in some ways akin to Jeremiah's promise of the new covenant written on men's hearts. But Ezekiel's vision of the future includes the picture of a holy nation, with the Temple of the holy God at the heart of its life (40–48).

The Fall of Babylon

In little more than forty years after Ezekiel's last recorded utterance, one part of the miracle had come about. Nebu-chadrezzar's long reign ended in 562. His son, Amel-marduk (the Evil-merodach who is referred to in 2 Kings 25^{27-30} as having released Jehoiachin) was king for only two years. After the brief reigns of two other kings, power was seized in 556 by Nabuna'id or Nabonidus, the last ruler of the neo-Babylonian empire. Though not himself a weakling, he weakened the empire and hastened its end. He spent some years at the oasis of Teima in the Arabian desert, leaving his son Belshazzar in charge of affairs in Babylon. One effect of the king's absence from the capital was that the annual New Year festival was not celebrated. By these and other actions he earned the hatred of the priests of Marduk. Consequently, it was a divided Baby-lon which had to face the rise of a new great power in western Asia.

That power was the empire of Cyrus the Persian. Cyrus had been a vassal of the Median king Astyages; but he revolted successfully against his overlord and gained control of the Median empire, which he forthwith proceeded to extend by a series of lightning conquests. A triple alliance was formed against him by Nabonidus, Croesus king of Lydia, and Amasis of Egypt. But Cyrus speedily invaded and occupied Lydia (546) and with Egypt in no position to offer effective help, Babylon was left to face the Persian assault alone. In 539 the city surren-dered to the Persians. It is significant that, in the record of the event, Cyrus claimed that he came by the command of the

Babylonian god Marduk, whom Nabonidus had failed to worship.[1]

Deutero-Isaiah

As, in the past, there had been prophets to interpret contemporary events in terms of Yahweh's chastisement of His people, so now, among the Babylonian exiles, there arose one who saw in Cyrus Yahweh's instrument for the deliverance of Israel. This was the unknown prophet whose poems are contained in Isa. 40–55, and who is usually referred to as the Second Isaiah, or, Deutero-Isaiah. These chapters contain no prose narrative; and we have no direct and explicit information about the prophet himself, though inferences have sometimes been drawn from parts of his prophecy.[2] But it is clear that the situation which he presupposes is that of the exiled Jews just after the middle of the sixth century. He looks forward to Babylon's downfall, the return of the exiles, the restoration of the homeland and the rebuilding of Jerusalem, and the spread of the knowledge of Yahweh among the nations.

Deutero-Isaiah sees Yahweh as Lord of history. Events which might be taken as just another turn in the game of power politics, or just another instance of the transience of human dominion and glory, are for him the furtherance of Yahweh's purpose and the fulfilment of His promise (40[8]). But to this central theme in prophetic teaching he gives a special application, which was relevant to the contemporary situation. Divination played an important part in Babylonian religion. With biting scorn Deutero-Isaiah challenges the pagan deities and diviners to show that they have been able to foretell what is now happening (41[21-24], 44[7], 45[21]). The God of Israel alone could foretell it, because He alone controls events (41[1-4]).

---

[1] See *ANE*, pp. 203 f., 206–8; *DOTT*, pp. 81–83.

[2] e.g., 49[1-6], where the first person pronoun is used. It has also been held that 52[13]–53[12] is a portrait of the prophet. The prophet does not even state directly where his ministry was exercised. The view assumed above, that he lived among the exiles, is generally held. But it has been argued that he lived in Palestine.

The prophet also sees Yahweh as Lord of nature, and as Creator of the universe (40^{12-31}). Here, again, he presents old truth with a new emphasis and in a more spacious context; and it has often been claimed that he is the first of the prophets to express an unqualified monotheism. It is, however, note-worthy that he is not concerned to deny in abstract terms the existence of other deities but rather to show their utter ineffec-tiveness ($41^{23 \text{ f.}}$). His scorn for idolatry is reminiscent of Elijah's mockery of the prophets of Baal ($40^{19 \text{ f.}}$, 44^{9-20}, 46^{1-7}).

This majestic God is concerned to meet the need of the exiled community (40^{27-31}). He is not only Creator of the universe, but Creator and Redeemer of Israel ($43^{1 \text{ ff.}}$). As of old He brought their fathers out of Egypt and across the sea, so now, in a new Exodus, He will bring them out of Babylon and across the desert to the promised land (43^{16-19}). In 40–48, the emphasis is laid on the deliverance from Babylon; but in 49–55 the thought is rather of the restoration of Zion, the rebuilding of the city and the repopulation of the land. Zion is described as the bride of Yahweh, temporarily separated from her Husband and bereft of her children, to whom her lost felicity would soon be restored (49^{14-26}, 54).

As the first Isaiah had seen Assyria as the rod of Yahweh's anger, Deutero-Isaiah saw Cyrus as the appointed instrument of His work of deliverance, Yahweh's 'shepherd' and His 'anointed' (44^{28}, 45^{1}). He even dared to believe that Cyrus would come to recognize Yahweh (45^{1-7}), though we have seen that when Babylon surrendered to Cyrus, he claimed to come by command of Marduk. It is sometimes suggested that it was disappointment at this turn of events which led the prophet to describe another figure: that of the Servant of Yahweh.

The Servant first appears in 42^{1-4}, a passage which some scholars have interpreted as a description of Cyrus, but which should probably be taken as a deliberate contrast to the picture of Cyrus in 41^{1-3}. The Servant's call, frustation, sufferings, death, and ultimate triumph are further described in $49^{1-6 \, (7)}$, $50^{4-9 \, (11)}$, $52^{13}-53^{12}$. These four Servant Songs present the supreme problem in Old Testament prophecy: Who was the

Suffering Servant? It is impossible here even to enumerate all the answers which have been offered to this question, still less to state and evaluate the arguments. The main *types* of solution are these: (*a*) corporate; i.e. the Servant is the actual Israel, or the ideal Israel, or a righteous nucleus in Israel; (*b*) individual; a historical figure from the past (e.g. Moses, Jehoiachin, Jeremiah), or from the prophet's own generation (e.g. Deutero-Isaiah himself, or a religious teacher whom he revered), or a future figure, such as the Messiah; (*c*) a combination of the corporate and the individual; i.e. the prophet's thought moved between the suffering and vocation of the community and those of an outstanding individual in whom Israel's suffering and vocation were to be supremely exemplified and realized.[1] Outside these four Servant Songs, the prophet explicitly identifies Israel with the Servant of Yahweh (41[8 ff.], 42[18 ff.]); and the same equation appears to be made in the second Song (49[3]).[2] But there are distinct differences between the portrait of the Servant in the Songs and that of the Servant-Israel elsewhere in the prophecy: e.g. the Servant-Israel is the recipient of salvation, whereas the Servant of the Songs is the mediator of it; and, again, the Servant of the Songs is said to have a mission to Israel. On the other hand, there is no individual historical figure of the prophet's own time or any earlier age with whom the exalted mission ascribed to the Servant could reasonably be associated. Accordingly the choice seems to lie between some form of corporate interpretation, a *future* individual interpretation, and a blend of the two.

It is clear that the portrait of the Servant incorporates many elements from Israel's past religious experience, and not least from the prophetic experience (e.g., cf. Isa. 49[1] with Jer. 1[5]). It has been claimed that there are also royal features in the portrait, or even that it is predominantly based on the religious

[1] This third line of interpretation is best understood in terms of the notion of corporate personality (see above, p. 15).

[2] On quite inadequate textual grounds, it has been suggested that the word 'Israel' should be deleted. Alternatively, 'Israel' is regarded as having an individual reference here.

functions of the monarchy. This view would suggest a Messianic interpretation of the Servant. But it rests in part on a debatable theory about the symbolic suffering of the king in the cult; and though some royal traits may be present in the portrait of the Servant, he is primarily a prophetic figure. Further, if the Servant is to be taken as an individual whom Deutero-Isaiah expected to come in the future, he is different from the Messiah as generally understood: he exercises no military or governmental authority, but is, rather, a mediator who establishes a right relationship between man and God.

In the figure of the Servant there are fused some of the great theological themes of the Old Testament. The experience of apparently unmerited suffering, which is described in moving terms in Jeremiah and Job and in many of the laments in the Psalter, is interpreted in the Servant Songs not only as a problem but as a vocation, by which God's will is accomplished through the Servant. The traditional Christian interpretation of the Songs has been that they are a prophecy of the ministry of Jesus of Nazareth, and, in particular, of His suffering and death. Sometimes this interpretation has been expressed in terms of a rather mechanical view of prophecy and of its fulfilment. But it is possible to reject such a view and yet to hold that the Songs found their fulfilment in Jesus. Even if we have to confess that the precise identification of the Servant which was present to the mind of the prophet eludes our inquiry, we may hold that his account of a divine ministry of reconciliation, faithfully and triumphantly exercised through suffering, was realized in Jesus.

The Servant's ministry reaches out beyond Israel ($42^{1, 4}$, 49^6). The thought that all men should be brought to worship and serve Yahweh is sometimes contrasted with passages in the prophecy which speak of the subjugation of other nations (43^3, 45^{14}, 49^{22-26}). The prophet was sufficiently aware of political realities to present the overthrow of other nations as steps towards the restoration of Israel; and he was sufficient of a poet to present this in highly coloured language without necessarily bringing it into strict logical accord with some other aspects of his teaching. But that he looked for the conversion of the nations to the God of Israel is clear from passages outside the Servant Songs. 'Turn to me and be saved, all the ends of the earth! For I am God and there is no other. . . . To me every knee shall bow, every tongue shall swear' ($45^{22\,f.}$; cf. 44^5).

Deutero-Isaiah shares neither the interest in ritual details nor the bizarre imagery which are both characteristic of Ezekiel's prophecies. But in his descriptions of the future he speaks not only of a historical deliverance but of a glorious transformation of the natural order (43^{19}, 51^{16}, $55^{12\,f.}$). In this he anticipates the promise of a new heaven and a new earth which was to feature prominently in later literature.

The Return: the Prophets of the Persian Period

Deutero-Isaiah's exultant prophecies of the restoration of national life in the homeland must have seemed strangely out of keeping with the actual experience of those who returned. For, although there were both the opportunity and the encouragement to return, conditions in and around Jerusalem were depressing, difficulties were many, and apathy was prevalent in the returned community. This much is clear and indisputable; but much else is obscure.

The period with which we are immediately concerned is the quarter-century following the fall of Babylon. Cyrus reigned until 530, when he was succeeded by his son, Cambyses, who invaded and conquered Egypt. When Cambyses died, leaving no direct heir, there was a struggle for power, the outcome of which was that Darius, son of Hystaspes, secured the throne; but it was only after further conflict that he was able to restore order in the vast empire.

The biblical evidence for events in Palestine during this period is contained in Ezra 1–6 and in Haggai and Zechariah 1–8. The two prophetic sources may safely be regarded as containing contemporary evidence about the situation which they describe. The chapters in Ezra belong to the Chronicler's history (1 and 2 Chronicles, Ezra, and Nehemiah), which can hardly be earlier than the fourth century; but they may well embody authentic material from an earlier time.

The opening verses of Ezra contain a decree of Cyrus, in Hebrew, granting permission for the return of the Jewish exiles, directing that the Temple at Jerusalem should be rebuilt, and ordering Jews who remained in exile to subscribe to the enterprise (1^{2-4}). Another version of this decree, in Aramaic, appears in Ezra 6^{3-5}, ordering that the Temple should be rebuilt, giving specifications for the rebuilding, and also providing for the return of the Temple vessels which Nebuchadrezzar had removed, but making no reference to a return of exiles. In both versions the date is given as the first year of the reign of Cyrus (i.e. 538, since the reference is clearly to his rule in Babylon). The story in Ezra 1 continues with an account of the return of a party of exiles, bringing the sacred vessels back to Jerusalem (vv. 7–11), under the supervision of 'Sheshbazzar the prince of Judah'. Who this Sheshbazzar was, we do not know; he is often identified with Shenazzar, a son of Jehoiachin (1 Chron. 3^{18}). It is stated that he was appointed governor and that he laid the foundations of the Temple (Ezra 5^{14-16}). Thereafter he disappears from history.

There is no need to doubt the substantial historicity of this account. The Persians showed considerable clemency in their treatment of subject peoples, and did not try to suppress local loyalties, customs, and cults. The Hebrew form of the decree (Ezra 1^{2-4}) admittedly displays some marks of the Chronicler's style. Presumably he recast the text to fit this part of his narrative. It has, however, often been questioned whether there was in fact any attempt to rebuild the Temple at this time. Ezra 3^{6-13} describes how the foundation of the Temple was laid in the time of Zerubbabel the governor (a grandson of Jehoiachin) and Jeshua the high priest, who were the leaders in the community in the time of Haggai and Zechariah; and the oracles in the book of Haggai, which are dated in the second year of Darius (i.e. 520 B.C.), presuppose that until then the Temple had remained in its ruined state. It is, however, neither difficult nor unreasonable to suppose that an abortive attempt at rebuilding was made by Sheshbazzar and his associates; and that the work was abandoned because of the depressing conditions which the returned exiles had to face.

The situation is briefly but vividly portrayed in the first oracle in the book of Haggai (1^{1-11}): drought, bad harvests, and what today would be called inflation ('he who earns wages earns wages to put them into a bag with holes'; 1^6). No doubt there were other embarrassments. Friction between the returned exiles and their neighbours appears to have begun early (Ezra 4^{1-5}); it was to recur with increasing sharpness in later generations. Any attempt to restore either the political importance of Jerusalem or the ancient glories of its Temple was liable to arouse jealousy and suspicion in other quarters; and those who, after the return from exile, regarded themselves as the custodians of the purity of Israel's religion, were apprehensive of corrupting influences from outside. Yet another factor in the situation in 520 was the recent sequence of events in the Persian empire. As, in the decade before the final fall of Jerusalem, the exiles had hoped in vain for the overthrow of Babylon, so now, the upheavals which followed the death of Cambyses prompted the expectation of a cataclysm which would restore Israel and the house of David to unchallenged pre-eminence.

In modern times Haggai is often described as pedestrian and prosaic. Nevertheless he had the ability to rouse a despondent and indifferent community to action. Asserting that their difficulties were a punishment for their failure to rebuild the Temple, he gave his hearers the assurance that, if they put first things first, prosperity would return (1^{3-11}). After the political upheavals of the time, the Temple, now a pitiful ruin of its former self, would be enriched by the tribute of the nations (2^{1-9});[1] and Zerubbabel would emerge as the chosen servant of Yahweh, His 'signet ring', giving effect to His

---

[1] The meaning of the oracle in 2^{10-14} is uncertain. It may mean that the people's indifference to the rebuilding of the Temple has polluted them, or, perhaps, that if they allow their half-pagan neighbours to join in the work, they will be polluted by them (cf. Ezra 4^{1-5}).

decisions; i.e. the heir of David (2^{20-23}). The rebuilding of the Temple and the restoration of the royal house are closely linked. Thus, in this short collection of prophecies, four important religious themes appear: (*a*) the close relationship between national religious faithfulness and prosperity which is an important element in the Deuteronomic teaching; (*b*) the positive attitude to the cult which is characteristic of post-exilic prophecy; (*c*) the expectation of a catastrophe which will be universal in its range (cf. 2^6), after which a new order will be established; and (*d*) the figure of Yahweh's royal representative as an element in the new order.

The same situation and similar aspirations are presented through the very different media of Zechariah's thought and style. The note of temporary disappointment is heard in the statement that 'all the earth remains at rest' (1^{11}); the revolts in the Persian empire have been suppressed; and the expected cataclysm is still in the future. Then, in eight dream-visions, the stages in the restoration of Israel are described, in imagery which is bizarre and at times difficult to interpret: the overthrow of Israel's enemies (1^{18-21}, 6^{1-8}); the rehabilitation and extension of Jerusalem (2^{1-5}); the divine vindication of Joshua, the high priest,[1] and the place which he and Zerubbabel have in Yahweh's care for the community (3, 4); the elimination of unworthy elements from the community (5^{1-4}); and the removal of Israel's sin (5^{5-11}). There is, in Zechariah, an echo of the old prophetic teaching, that if justice and compassion are neglected, ritual observances are worthless (7). But, like Haggai, he is insistent that the Temple must be rebuilt speedily (1^{16}, $4^{7, 9}$). For him, as for Haggai, Zerubbabel is the servant of Yahweh, the 'Branch' of David's line (3^8, 6^{12}; cf. Isa. 11^1; Jer. $23^{5\ f.}$, 33^{14-16}), but Joshua the high priest is more prominent here than in Haggai's oracles.[2] This is a pointer to

[1] 'Joshua the son of Jehozadak', who is referred to in the books of Haggai and Zechariah, is, of course, identical with the 'Jeshua' mentioned in Ezra $3^{2, 8}$, 4^3.

[2] In Zech. 6^{11}, the name 'Zerubbabel', which no doubt originally stood in the text, has been replaced by 'Joshua'.

the growing influence of the high priest in the affairs of the Palestinian Jewish community, of which, in time, he became the civil as well as the religious head.

Zechariah's prophecies contain a number of features which are characteristic of the apocalyptic literature of later times. Visionary experiences had played a part in the ministry of earlier prophets; but the sequence of elaborate dream-visions which are recorded by Zechariah dominate his recorded teaching. Moreover, they include artificial symbolism which is not characteristic of his prophetic predecessors (with the possible exception of Ezekiel) but is a constant mark of apocalyptic. Another such feature is the reference to angels, and, in particular, to an angel who interprets what the prophet sees (1^9 ff., 18 ff., 2^3 ff., etc.). The reference to 'Satan' (i.e. 'the Opposer'; see 3^{1} ff; cf. Job 1^{6-12}, 2^{1-7}; 1 Chron. 21^1) is not to be taken at this stage as indicating belief in a personal Devil, but is nevertheless an important presage of later developments in the conception of an arch-fiend.[1]

The prophecies of Haggai and Zechariah stimulated Zerubbabel and Joshua to organize the rebuilding of the Temple. The enterprise was questioned by the Persian satrap, Tattenai; but when he referred the matter to Darius, the decree of Cyrus was found among the royal records. Thereafter, the work went forward with official approval and support, until, in 515, the rebuilding was completed (Ezra 5, 6).

For the next half-century or so there is no indisputable evidence about the condition of the Jewish community in Palestine. Zerubbabel disappears from the scene, leaving behind him no line of Davidic kings, but only the restored Temple as the legacy of his leadership. Probably the community lapsed into despondency and religious indifference. This, at all events, is the picture given in the book of Malachi, which reflects conditions shortly before the middle of the fifth century. Temple worship was continued in a slovenly and half-

---

[1] The account of the teaching of Zechariah which is given above is based entirely on chapters 1–8. It is generally recognized that chapters 9–14 are of later origin.

hearted fashion (Mal. 1^{6-14}, 3^{6-12}); the priests were casual in the fulfilment of their responsibilities as teachers (2^{1-9}); Jewish wives were divorced so that their husbands might contract marriages with foreign women (2^{10-16}); and a mood of despondent cynicism and practical atheism was prevalent ($3^{13 \, f.}$). Malachi's[1] message in this situation was a blend of rebuke and appeal. By contrast with those pre-exilic prophets who had denounced the cultic practice of their countrymen as an offence to Yahweh, he condemned neglect of the appointed ways of worship and offering; and in this way his teaching exemplifies the growing emphasis on the Temple and the law; but the tone and spirit of his oracles are strikingly akin to pre-exilic prophecy. Unlike Haggai and Zechariah, he did not see among his contemporaries any outstanding figure who was the chosen of Yahweh, but only a group of faithful Jews, whom Yahweh would acknowledge as His own, when He came in judgement (3^{16-18}).

Ezra and Nehemiah

The Jewish community which lived in and around Jerusalem had benefited considerably from the relatively enlightened policy of the Persian rulers. The return of exiles had been sanctioned;[2] and the rebuilding of the Temple had been permitted and subsidized. But serious difficulties remained: material conditions were depressing; the Jews were surrounded by jealous and suspicious neighbours; and they themselves were prone to religious indifference. The concluding sections of the Chronicler's history (Ezra 7–10; Nehemiah) tell of the work of two men, Ezra and Nehemiah, who came to Jerusalem, the one to regulate the life and worship of the community

---

[1] 'Malachi' means 'my messenger'; and it may well be that it was not the personal name of the author of the prophecies, but was derived from 3^1 and applied as a kind of sobriquet to the anonymous prophet.

[2] 'The return' was not a once-for-all event. Parties of Jews migrated back to the homeland at different times. It may be that Zerubbabel and Joshua were the leaders of such a party. The narratives about Ezra and Nehemiah provide evidence of later migrations. No doubt there were others of which no evidence has survived.

according to 'the law', the other to rebuild the ruined fortifications of the city and to provide it with stable administration
and security. Unfortunately, the narrative raises perplexing
problems, the most important of which for our present purpose
is chronological.

The Chronicler appears to date Ezra's arrival in Jerusalem
a few years earlier than that of Nehemiah, and to regard the
two men as contemporaries whose work in Jerusalem overlapped. Ezra came to Jerusalem in the seventh year of
Artaxerxes (Ezra 7[7]), and Nehemiah in the twentieth year
of Artaxerxes (Neh. 2[1]). After Nehemiah had rebuilt the walls
and secured the orderly life of the community, Ezra and his
assistants conducted a public reading of the law, which was
followed by a celebration of the feast of Tabernacles (Neh. 8,
9). In the thirty-second year of Artaxerxes, Nehemiah returned
to the Persian court, but later came back to Jerusalem (Neh.
13[6 f.]). If, as was generally assumed until the end of the nineteenth century, the Persian king referred to is Artaxerxes I
(465–424), then Ezra's journey is dated in 458/7, and Nehemiah's first arrival in 445/4 and his second some time after 433.
But on various grounds it has been argued that Ezra's arrival
must have been considerably later than 458. Some scholars
hold that he came later in the reign of Artaxerxes I: on this
view it is necessary to suppose that the reference to 'the seventh
year' in Ezra 7[7] is a scribal error, and that some other numeral
should be substituted, e.g. 'the twenty-seventh', which would
bring Ezra's mission down to 428. Others maintain that Ezra's
work was done during the reign of Artaxerxes II (404–358),
and that he came to Jerusalem in 398/7. The arguments in
support of these views are not wholly conclusive. The whole
question is complicated by problems about the Chronicler's
use of his sources, possible additions to his history by later
hands, and differences between the Hebrew and Aramaic
text on the one hand[1] and the text of the Greek Ezra (I Esdras
in the Apocrypha) on the other. Every theory of the chronology is beset by difficulties; and while the date of Nehemiah's

[1] Ezra 4[8]–6[18] is in Aramaic.

work is virtually certain, any date to which Ezra's mission is assigned must, in the present state of our knowledge, be provisional. A brief summary of the arguments for a later dating of Ezra is given in a separate note at the end of this chapter.

The narrative contains autobiographical passages relating to both Ezra and Nehemiah. The latter are more extensive (Neh. 1^1–7^{73a}, 11–13) and do not bear the marks of editing by the Chronicler which are evident in the Ezra memoirs. All in all, the picture of Nehemiah which has been preserved is bolder in its outlines and livelier in its colouring than that of Ezra the scribe. Nehemiah was a high official at the court of Artaxerxes, who, hearing of the ruined state of the fortifications of Jerusalem and the plight of the Jews there, asked and received the king's authority to direct the work of rebuilding. It is uncertain whether the ruined state of the walls was the result of recent incidents. Ezra 4^{17-22} records how, during the same reign, an attempt at rebuilding was stopped by royal decree. At all events, Nehemiah set about the task with discretion and vigour, and so organized the work and inspired the workers that the wall was completed in fifty-two days. (Neh. 6^{15}). This was done in the face of external opposition, headed by Sanballat the Horonite (usually identified with the Sanballat who is mentioned in the Elephantine papyri as governor of Samaria $c.$ 410), Tobiah the Ammonite, and Geshem the Arabian. After the fall of Jerusalem and the dispersion of the community at Mizpah, Judah and Jerusalem had probably come under the administrative control of Samaria. It was, therefore, only natural that jealousy and suspicion should be aroused in that quarter when a resolute governor appeared in Jerusalem who was determined to restore to it something of its ancient strength. Ridicule, fifth-column activity, misrepresentation, threats, and attempted assassination were all employed in the effort to bring Nehemiah's enterprise to an end (Neh. 4, 6). But he was too shrewd and courageous to be outwitted or intimidated.

When the wall had been completed, Nehemiah increased the depleted manpower of the city by drafting into it a tenth of the population of neighbouring districts ($7^{4,\ 5a}$, $11^{1\ f.}$). There were also abuses to be remedied. Prolonged economic difficulties and heavy taxation caused great hardship to the poorer classes. As in the eighth century, callous creditors had foreclosed mortgages, depriving debtors of their land and even reducing their children to slavery. Nehemiah made the offending upper classes promise solemnly to abandon the practice of usury, and to restore the estates which had been appropriated (5^{1-13}). He himself did what he could to alleviate the financial burdens on the community by waiving his claim to some of the customary emoluments of his office (5^{14-19}).

Nehemiah's patriotism was inseparable from his piety. The shrewd and energetic rebuilder of the walls of Zion would turn in prayer to the God of his fathers in the midst of his perplexities and dangers (2^4, $4^{4,\ 9}$, 6^9); and the efficient organizer of the affairs of the community had a concern for its religious practice and purity. He brought back the Levites from secular employment to their proper duties and regulated the payment and administration of tithes (13^{10-14}). He enforced a stricter observance of the Sabbath (13^{15-22}). One of the chief threats to religious strictness came from foreign influence; and one of the chief channels of foreign influence was the priesthood. The high priest Eliashib had connexions with Nehemiah's adversary, Tobiah, for whom he provided accommodation in the Temple, during the interval between Nehemiah's two terms as governor. On his return, Nehemiah threw out Tobiah's belongings, and had the room purified and restored to its proper use (13^{4-9}). Incensed at the widespread infiltration of alien elements, Nehemiah exacted a pledge that intermarriage with foreigners should cease (13^{23-27}). A grandson of Eliashib, who had married Sanballat's daughter, was sent packing ($13^{28\ f.}$). It has been said that in Nehemiah's policy lie the beginnings of Jewish exclusiveness: it can be said with equal truth that in the circumstances of his time we may see the justification of it.

It would be a strange historical irony if Nehemiah and Ezra were contemporaries and colleagues: the vigorous man of action, who prayed to God and set a guard against a possible assault (Neh. 4^9), and the priest, who was ashamed to ask for a military escort, lest it should suggest lack of faith in God (Ezra 8^{22}). But, as we have seen, we cannot be certain of their chronological relationship. If Ezra came to Jerusalem in 458/7, then he appears to have waited a dozen years or more before promulgating the law (Neh. 8, 9). To account for the delay, it has been suggested that Ezra's work became possible only when Nehemiah had established security and stability, or that Ezra may at first have been discredited because of his rigorism. On any alternative dating, it must be assumed that Neh. 8 and 9 are misplaced. If Ezra came to Jerusalem at a later date in the reign of Artaxerxes I (e.g. in 428), then he did his work during Nehemiah's second term as governor, which seems to have been the time when most of Nehemiah's religious measures were enforced; and it is something of a difficulty that Ezra is not mentioned in connexion with any of them. If, however, Ezra's mission is dated in the reign of Artaxerxes II (i.e. in 398/7), the work of the two men was entirely independent, except that Ezra must have benefited from those results of Nehemiah's work which had survived the intervening decades.

Ezra's title, 'the scribe of the law of the God of heaven' (7^{12}), was an official Persian one, indicating that he came as a commissioner for religious affairs among the Jews in the province. He had official authorization and financial backing, and was accompanied by a considerable number of Jews. He also brought with him 'the book of the law of Moses' (Neh. 8^1; cf. Ezra 7$^{6, 10, 14}$). At a solemn assembly of the people, Ezra and his assistants read the law aloud, translating it into Aramaic to ensure that it should be understood (Neh. 8^8).[1]

We do not know the identity of the law book which Ezra

---

[1] In post-exilic times Aramaic became the current language of Palestine. Neh. 8^8 has been traditionally regarded (probably rightly) as recording the beginning of the practice of following the reading of a portion of the Hebrew Scriptures in the Synagogue by a translation or paraphrase into Aramaic. Such translations, originally oral but subsequently written, are known as Targums or Targumin.

brought with him. Deuteronomy, part or all of the Priestly Code (cf. above pp. 32, 36) and the entire Pentateuch have all been suggested. The first of these is very unlikely. In the account of the celebration of the feast of Tabernacles which followed the public reading of the law, the details agree with the Priestly Code as against Deuteronomy, in that (a) the date of the feast is fixed at a point in the calendar, (b) the length of the festival is eight days, not seven, and (c) there is evidence of the explicit injunction to live in booths during the period of the feast (cf. Neh. 8¹³⁻¹⁸ with Lev. 23³³⁻⁴⁴, and contrast with Deut. 16¹³⁻¹⁷). But, whether we date Ezra as early as 458/7 or as late as 398/7, there are fairly strong grounds for holding that by his time the main components of the Pentateuch had been brought together. The Pentateuch is the Bible of the Samaritans; and although it is impossible to establish with accuracy or certainty the date at which the decisive break between Jews and Samaritans occurred, it is unlikely to have been later than the time of Alexander the Great. A holy book which is the common possession of both communities must surely have been accepted by both for a considerable time before the period of acute hostility which led to the break. But although, as seems likely, Ezra's law book was the entire Pentateuch, complete in substance though not in its final textual form, there is nothing in the narratives to suggest that it was being introduced as a new law. Ezra's mission seems rather to have been to enforce authoritatively the provisions of a book which had been in existence for some time. Whatever differences there may have been between the promulgation of the book found in the Temple in Josiah's time and the public reading of the law after Ezra's arrival in Jerusalem, there is at least this similarity, that on both occasions a book was recognized as authoritative and regulative for the life of the community. Zerubbabel and Joshua had given the restored Jewish community a restored sanctuary as the centre of its religious life. Ezra established as normative a sacred book which provided directions for worship, for the functions of the hierarchy, and for the general religious practice of the community. This was a step of great importance in the development in Judaism of the authority of the law (*Torah*), a feature of Israel's religious heritage which remained when, once again, the Temple was destroyed and sacrifice could no longer be offered.

The dominant legislative element in the completed Pentateuch is the Priestly Code. Not that the Priestly material is entirely legal; but even the narratives which are written in the Priestly style are usually related to some feature of religious observance (e.g. the story of creation ends with an impressive reference to the law of the Sabbath; Gen. 2$^{1-3}$). The mass of detailed regulations relating to ritual purity and cultic practice reflects a concern to preserve free from blemish or deviation the worship of the God of Israel, an aim which the Deuteronomic reformers had sought to attain by the law of centralization. The work of codifying these detailed regulations had probably been carried out in large measure during and after the Exile; but much of the material so codified may well have been much older. It is impossible to draw a sharp line of demarcation between what was old and what was new; but certain features may be noted which are absent from earlier codes. The system of festivals and other sacred occasions is presented in greater detail and dated precisely by the days and months of the calendar and not merely in relation to the phases of the agricultural year (Lev. 23). It is only in the Priestly Code that we find the regulations for the Day of Atonement (Lev. 16), and for the sin offering and guilt offering (Lev. 4–6^7). The priestly order neither includes all Levites nor (as in Ezekiel) excludes all but Zadokites, but consists of the house of Aaron (Exod. 28^1). The other Levites perform subordinate functions. The high priest is the supreme figure, not only in the hierarchy, but also in the entire community, embodying some of the features of the monarchy. He presides over a community whose separateness, ritual purity, and ethical intregity it is the aim of the law to secure.

The maintenance of a pure religious tradition was made difficult by contact with foreigners, and, in particular, by intermarriage with them. Malachi had denounced the practice of divorcing Jewish wives in order to marry aliens (Mal. 2^{10-16}). Nehemiah imposed a promise that intermarriage with aliens should cease (Neh. 13^{23-27}). Under Ezra a more drastic step was taken: Jews who had married foreign women were required to divorce them (Ezra 9, 10). In considering these stringent measures, and other marks of exclusiveness in the Judaism of this and later periods, two points need to be borne in mind. The first is that the aim was the maintenance of *religious* purity, and that the erecting of racial or national barriers were measures to that end. The second is that the dangers of assimilation were real and serious. The Jewish community in and around Jerusalem was small and insecure and was subject to powerful and insidious alien pressures. It is noteworthy that the priestly families were particularly susceptible to such influences.

The book of Ruth has often been regarded as a direct criticism of the opposition of Ezra and Nehemiah to mixed marriages. It tells how a woman from Moab married into a Judahite family and became the ancestress of King David. But the story has, in fact, none of the characteristics of a polemical tract, and is more likely to have been written to show the guiding hand of God in the lives of David's forebears. On the other hand, the book of Jonah is a telling plea against attempts to regard even the most hated of foreign nations as excluded from the mercy of God. As such, it presses home with emphasis and concentration the concern, already present in Deutero-Isaiah, that other nations should come to worship and obey the God of Israel. Running through much of the religious literature of the period from the Exile onwards there are these two contrasted features: on the one hand, the concern to preserve the purity of Israel's religious inheritance and the separateness of the Jewish community; on the other, the recognition that all nations have a place in the purpose of Israel's God. In modern books these seemingly antithetical tendencies are often referred to by the somewhat abstract terms 'universalism' and 'particularism'. However convenient such terminology may be, it should not be allowed to obscure the fact that both features had their source in Israel's religious tradition and were vitally related to the actual situations in which Jewish communities found themselves, in Palestine and throughout the Dispersion.

Deviations: The Samaritan Schism

In spite of the efforts of men like Nehemiah and Ezra to regulate Jewish life and worship by the requirements of the law, it is clear that even in Jerusalem itself, and even (perhaps particularly) among the ranks of the priesthood, there were recalcitrants; it would be wrong to suppose that further afield there was strict uniformity. The Elephantine papyri (see above, p. 141, n. 1) record in a fragmentary fashion aspects of the life of an interesting Jewish community in Egypt. One document, which unfortunately is badly mutilated, contains an order from the Persian king, Darius II, that the community should observe a festival which clearly is that of Unleavened Bread; and it may well be that a lost portion of the letter referred to the Passover. This is an important illustration of the active interest which the Persian rulers took in the religious practice of subject peoples. But the community at Elephantine did not conform to Deuteronomic standards. It had its own temple, at which worship was offered not only to Yahu (Yahweh) but also to other deities.[1] Yet, in spite of this breach of the Deuteronomic law of the single sanctuary, the leaders of the Elephantine community appealed to Bagoas, the governor of Judea, and to the priestly authorities in Jerusalem, for support in their attempt to get their temple rebuilt after it had been destroyed in anti-Jewish riots in 410 B.C. It is perhaps not surprising that this letter was left unanswered. In 407 B.C.

---

[1] The names Anath-yahu, Anath-bethel, Herem-bethel, and Ishum-bethel occur. It has been suggested that they refer not to independent deities but to 'aspects' of Yahweh which were regarded as having some measure of personal divine status. But, even on that view, the religion of the Elephantine Jews was far removed from assertions like 'The LORD our God is one LORD' (Deut. 6⁴), or, 'I am the LORD, and there is no other, besides me there is no God' (Isa. 45⁵).

further appeals were sent to Bagoas and to the authorities in Samaria, who both responded favourably.

It is interesting that the Jews of Elephantine not only looked to Palestine for support, but corresponded with both Jerusalem and Samaria. From this it has been inferred, on the one hand, that they tried to play the one off against the other, and, on the other hand, that no decisive breach had as yet occurred between Jews and Samaritans. No firm conclusion can be drawn from the evidence; but, whatever variant forms Judaism may have taken in the Dispersion, it is evident that in Palestine itself there were two distinct communities, Jews and Samaritans, both of which, either at this time or soon afterwards, acknowledged the authority of the five books of Moses. Before the New Testament period, both communites had not only separated, but had become bitterly hostile to each other. It is, however, impossible to determine precisely when the final breach took place, and difficult to trace the development of misunderstanding and enmity.

Some have tried to trace the origins of the schism to those geographical and tribal factors which contributed to the division of the kingdom in Rehoboam's reign. In the Old Testament itself, the Deuteronomistic historian follows his account of the fall of Samaria in 722 with a polemical account of the settlement in the northern territories of an alien population and the establishment of a corrupt but partly Israelite religious tradition (2 Kings 17^{24-41});[1] whereas the Chronicler is significantly silent about almost the entire history of the northern kingdom. It is clear that, after the Exile, the restored community in Judea did not enjoy easy relations with the inhabitants of the northern region, of which Samaria was the administrative centre. But the evidence is tantalizingly fragmentary. An offer of help in the rebuilding of the Temple was rebuffed by the Jews (Ezra 4^{1-5}; cf. Hag. 2^{10-14}; and see above,

---

[1] A very different Samaritan tradition traces the division back to the time of Eli, who is said to have sought the high priesthood for himself and to have established at Shiloh a rival sanctuary to the one on Mount Gerizim, the Samaritans' holy mountain.

p. 155, n. 1). In the reign of Artaxerxes I, the rebuilding of Jerusalem was obstructed (Ezra 4⁷⁻²³). Nehemiah's work in restoring the defences of Jerusalem was constantly threatened by the efforts of Sanballat and his associates (Neh. 4, 6). It is generally assumed that this is the Sanballat who is referred to as governor of Samaria in the Elephantine correspondence of 407 B.C. Behind such clashes lay a mixture of political and religious factors; and the situation was further complicated by personal and family relationships. Nehemiah's enemy, Tobiah, was befriended by the high priest Eliashib, one of whose grandsons married Sanballat's daughter and was expelled by Nehemiah (Neh. 13²⁸). The latter incident has sometimes been regarded as the occasion of the irrevocable breach between Jews and Samaritans; but there is nothing in the biblical text to suggest this. The Jewish historian Josephus (*Antiquities of the Jews*, XI, vii. 2; viii. 2 ff.) states that Manasseh, brother of the high priest Jaddua, was prevented from officiating as a priest at Jerusalem because he had married the daughter of Sanballat, the governor of Samaria, and that Sanballat built a Temple for him on Mount Gerizim. There are obvious points of similarity here with the story in Nehemiah; but there are also differences; and Josephus dates these events a century later, in the time of Alexander the Great. It is impossible to decide how much of historical fact there is in Josephus's narrative, or whether he is right in assigning the origin of the Samaritan Temple to that date. In any event, the actual building of the Temple was less important than the Samaritan claim that Mount Gerizim, and not Mount Zion, was the divinely appointed centre of worship (John 4²⁰). In the Samaritan text of the Pentateuch, the Decalogue ends (Exod. 20¹⁷; Deut. 5²¹) with a command to sacrifice on Mount Gerizim; and in Deut. 27⁴, where the Jewish text has Mount Ebal, the Samaritan text has Mount Gerizim as the place for the building of an altar and the inscribing of the law. These rival claims for Mount Zion and Mount Gerizim constituted the basic religious difference on which centred the intense hostility between the two communities.

From Alexander the Great to the Maccabaean Revolt

During the first half of the fourth century the Persian empire had been progressively weakened by a succession of revolts and palace intrigues. When Darius III (336–331) came to the throne, Persia was in no condition to withstand the eastward thrust of the armies of Alexander the Great (336–323). A victory at the river Granicus gave Alexander control of Asia Minor; and a year later the battle of Issus opened the way to Syria, Palestine, and Egypt, into which he advanced before proceeding further east. At Gaugamela (331) he struck his final blow. The power of Persia was at an end; and Alexander's seemed to be just beginning. His almost incredible advance into India looked like the prelude to a period of imperial consolidation on an unprecedented scale. But in 323 Alexander died; and his empire became the prey of rival generals. For the remainder of the Old Testament period the history of Palestine was largely affected by the policies of royal houses founded by two of these generals: the house of Ptolemy in Egypt, and the house of Seleucus, who gained control of a large part of western Asia including Syria. As had happened so often in the past, Palestine was both disputed territory and a corridor between two rival powers. During practically the whole of the third century it was controlled by the Ptolemaic rulers of Egypt, from whom it was finally wrested by Antiochus III (the Great) of the house of Seleucus. His first attempt was foiled by an unexpected Egyptian victory at Raphia, in the extreme south-west of Palestine (217). But in 198, at Panium near the sources of the Jordan, he decisively defeated Egypt and added Palestine to his empire.

Throughout this entire period the Jews, like the other inhabitants of the oriental regions ruled by Alexander's successors, were exposed to Greek cultural and social influences. New cities were built and organized on the Greek model; and older cities were transformed to the same likeness. On many sites throughout the Near East today the remains of theatres, stadia, and hippodromes are reminders of this process of westernization. The Greek language was increasingly used as a medium of communication. As, during the Persian period, Aramaic had become the daily speech of the Jews in Palestine, Babylonia, and elsewhere (cf. above, p. 162, n. 1), so now, Greek became the vernacular in many Jewish communities. Large numbers of Jews had settled in Egypt; and in the great new city of Alexandria there was a particularly influential community. According to ancient tradition, as preserved in a document known as *The Letter of Aristeas*, it was there that the first translation of the Old Testament into Greek was made: the Septuagint (LXX), so called because seventy-two elders are said to have produced it. Ptolemy II Philadelphus (285–247) is represented as having been the patron of the enterprise. This account of the making of the translation is probably in large measure unhistorical; but it does point to the general situation and period in which a Greek translation of at least the Pentateuch was made. If Ptolemy II did not in fact play the active role in this undertaking which is assigned to him, the Ptolemaic rulers appear, on the whole, to have given tolerant and considerate treatment to the Jews in Egypt and Palestine. But reliable evidence for the period is meagre.

There was no immediate change for the worse after the battle of Panium. Seleucid control of Palestine lasted for about a quarter of a century before any really serious crisis arose. When Seleucus IV (187–175), the successor of Antiochus the Great, was assassinated by his minister Heliodorus, power was seized by Antiochus IV Epiphanes (175–163), in whose reign loyal Jews had to face a major threat to their existence as a religious community.

Complex factors contributed to produce the crisis and to shape the conflict which followed. For a century and a half, even under tolerant rule, the Jews had been subject to the pressure of Greek influence. Many, particularly among the aristocratic and priestly classes, had welcomed it and had readily conformed to Greek ways; others resisted alien influence, seeing in it a threat to their religious integrity. There was, therefore, no question of a direct clash between 'the Jews' and their rulers; for the Jews were divided amongst themselves. As for Antiochus, his policy was not simply an expression of anti-Jewish feeling, whether racial or religious, but was affected partly by the need to hold together the multifarious elements in his empire and to consolidate his power against Parthia, Egypt, and Rome, partly by his financial straits, and partly by his own vehement and unbalanced personality. He sought to impose upon his subjects the unifying bond of Greek culture, and of a blend of Greek and oriental cults. His title, Epiphanes, expresses his claim to be the earthly manifestation of Zeus Olympius, a claim which is reflected in the representation of Antiochus on his coins.

Corruption in the ranks of the Jewish priesthood, and rivalry for the supreme position in the hierarchy, made it relatively easy for Antiochus at one stroke to increase his influence in Jerusalem and replenish his resources. During the absence of the legitimate high priest, Onias III, his brother Jason (i.e. Joshua) brought about the deposition of Onias and his own appointment, by bribing Antiochus and promising to further the Hellenization of Jerusalem (2 Macc. 4[7 ff.]; cf. 1 Macc. 1[13-15]). Three years later Jason was himself deposed when a certain Menelaus bought the office for a larger sum (2 Macc. 4[23 ff.]). In 169, encouraged by a false rumour that Antiochus had been killed in Egypt, Jason returned to Jerusalem, drove out Menelaus, and re-established himself as high priest. The situation was soon reversed once more. On his way back from Egypt, Antiochus restored Menelaus and plundered the Temple (1 Macc. 1[20-24]; 2 Macc. 5[11-21]). But opposition to foreign domination and to the king's Hellenizing policy was growing. In 167 Antiochus was obliged to send to Jerusalem a substantial force under Apollonius, one of his generals. Many Jews were treacherously put to death, the city was plundered and its fortifications destroyed, and a Syrian stronghold (the Acra) was established within the confines of Jerusalem (1 Macc. 1[29-35]; 2 Macc. 5[24-26]). The king's next move was a blow at religious practice: circumcision, the observance of the Sabbath, and of the festivals, and the Jewish sacrifices were forbidden;

copies of the law were destroyed and the possession of them was made a capital offence; pagan altars were put up in various parts of the country; the cult of Zeus Olympius was introduced into the Temple, in December 167, where an altar to Zeus was built over the altar of sacrifice and perhaps also an image of him erected (1 Macc. 1⁴¹⁻⁶⁴).[1] In spite of the ruthlessness with which the royal decree was enforced, there was considerable Jewish resistance. The group known as the Hasidim (i.e. the pious ones), who were the spiritual ancestors of the Pharisees, the Essenes, and the Qumran sect, displayed unswerving loyalty to the law. But it was evident that passive disobedience to the king's decree was not enough. Active rebellion broke out at Modein when a priest named Mattathias not only refused to offer pagan sacrifice, but killed both an apostate Jew and also the officer who was enforcing the decree (1 Macc. 2¹⁵⁻²⁶). He and his five sons then took to the hills, where they were joined by others including Hasidim, who were determined to take the offensive against their oppressors. After the death of Mattathias in 166, the leadership passed to his third son Judas, nicknamed Maccabaeus (the Hammer), who won a series of victories over the Syrian armies, culminating in the occupation of Jerusalem. The Acra remained in Syrian hands; but it was now possible to cleanse the Temple and to restore legitimate worship there exactly three years after the act of profanation (December, 164). This has been celebrated ever since in the feast of Hanukkah (Dedication). Notable as the achievement was, the campaigns of Judas and his brothers were by no means over; but to pursue the subsequent course of events would take us beyond the Old Testament period.

The Book of Daniel and the Rise of Apocalyptic Literature

The book of Daniel reflects both the sharp challenge of Seleucid persecution and the subtler pressures of Greek

---

[1] The term 'the abomination that makes desolate' (Dan. 11³¹, 12¹¹; cf. 1 Macc. 1⁵⁴) is a punning distortion of 'the Baal of Heaven', the Semitic equivalent of Zeus Olympius.

influence which had preceded it. Though it purports to describe the trials of four young Jewish noblemen in the conditions of the Babylonian Exile and the beginning of the Persian period, it relates directly to the situation of faithful Jews in Palestine during the reign of Antiochus Epiphanes. In the stories of Daniel and his companions refusing to eat forbidden food, to worship the great idol, and to pray only to the king (Dan. 1, 3, 6), we may recognize the tests imposed by the Syrian authorities. The den of lions and the burning fiery furnace are apt expressions of the savage persecution which the pious had to endure; and the words of Daniel's companions to Nebuchadnezzar might well be taken as the confession of faith of the Hasidim: 'If it be so, our God whom we serve is able to deliver us from the burning fiery furnace; and he will deliver us out of your hand, O king. But if not, be it known unto you, O king, that we will not serve your gods or worship the golden image which you have set up' ($3^{17\,f.}$). These stories in the first half of the book are an exhortation to the persecuted Jews to remain faithful, and the expression of a rock-like faith in the supremacy of the God of Israel over the power of Antiochus, represented by the tyrants in the stories.

Combined with this there is a view of history from the Exile to the time of Antiochus Epiphanes. Four successive world empires dominate the four phases into which it is divided; but in a final crisis God will overthrow the power of evil, and the trials of His faithful people will be at an end. This appears already in the dream of King Nebuchadnezzar (Dan. 2). It is further developed throughout the second half of the book, in the fantastic imagery of the four beasts (i.e. four empires) in 7, and in the elaborate symbolism of other passages. The climax of the entire process is at hand, in which the bestial empires will be superseded by 'one like a son of man', i.e. 'the kingdom of the saints of the Most High' ($7^{13\,f.}$, 27).[1] When evil has finally

[1] Elsewhere in the Old Testament (e.g. in Ps. 8^4; Ezek. 2^1, and many other passages in Ezekiel) the term 'son of man' means 'man', 'mere man', 'mortal man'. Here in Daniel it is used as the title of a figure who represents the community of saints. But in later apocalyptic writings (e.g. 1 Enoch;

been overthrown, there will be a resurrection of the righteous to a glorious life, and of the wicked to punishment ($12^{2\,f.}$).

Daniel displays clearly the main features of apocalyptic, a type of literature which played an important part in Jewish religion between the Maccabaean period and the end of the first Christian century. In the Old Testament itself there is no other fully developed apocalyptic work; but in the transformation through which prophetic literature passed from the sixth century onwards it is possible to see important lines of development in the direction of apocalyptic. We have already noted the occurrence in Ezekiel and in Zech. 1–8 of significant characteristics: fantastic and artificial symbolism, dream visions, angelic interpreters, and the like. Other indications of the development towards apocalyptic may be seen in the prophecies about Gog in Ezek. 38, 39; in the puzzling collection of hymns and prophecies in Isa. 24–27, sometimes called 'the Isaiah Apocalypse'; and in Zech. 9–14. In the earlier period, prophets sought to convey their message by public speech and action; only subsequently was it written down, so that it might be preserved and communicated to readers as well as hearers. But in the later period (i.e. from about the end of the monarchy onwards) prophets seem increasingly to have used the written word. Thus prophecy became consciously literary. Apocalyptic may be regarded as the culmination of this development. It is a learned literature, in which earlier prophecies are sometimes quoted and reinterpreted. Further, we find in it, not the relatively simple and obvious imagery of the earlier prophets, but, as we have already seen, artificially constructed symbolic images, such as the zoological monstrosities mentioned in Dan. 7^{1-8}. In keeping with this is the symbolic use of numbers. Again, whereas many prophetic oracles are anonymous, apocalyptic writings are nearly always pseudonymous, written in the name of some ancient worthy such as Daniel or Enoch. Various explanations of this characteristic have been

2 Esdras) the 'Son of Man' is an individual, pre-existent, heavenly being. These books (including Daniel) are an important part of the background of the title as applied to Jesus in the Gospels.

offered: e.g. that, since the age of inspiration was thought to have ceased, the claim of antiquity was necessary to guarantee the authority of these writings; or that, in accordance with the notion of corporate personality (see above, pp. 15, 151), the apocalyptist thought of the teaching of the ancient seer as continued in himself; or that the convention began with the author of the book of Daniel, who, having written stories about Daniel, recorded his visions in the name of Daniel, simply to indicate that the stories and visions came from the same author. Whatever cause or combination of causes may have given rise to the element of pseudonymity, literary convention no doubt ensured its continuance.

The word 'apocalypse' means 'revelation' or 'unveiling'. The truth which the apocalyptists sought to unfold was a secret knowledge, imparted only to the faithful. There are descriptions of the course of history, leading up to the final crisis, presented as visions of the ancient seers, or, sometimes, as their *testaments* to their descendants: here, what is presented as future is mostly past or contemporary from the standpoint of the apocalyptists and their readers. Some apocalypses tell not only of the progress and climax of history, but of how the seer was allowed to visit the heavenly regions.

In all of this there is ample use of ancient mythological imagery. Many scholars have also claimed that Persian influence is evident in the apocalyptic teaching about the division of history into ages, the conflict between the powers of good and evil, the hierarchies of angels and devils, the final decisive crisis, and the resurrection of the dead. That there was some Persian influence on post-exilic Judaism, and on apocalyptic in particular, it would be rash to deny; but the main features of apocalyptic teaching are developments and elaborations of elements which were already present in the religion of Israel. The apocalyptic message, as we see it in Daniel and similar books, was addressed to faithful Jews in times of persecution and crisis, assuring them that the Day of the Lord was at hand, the decisive crisis, in which His righteous rule would once and for all be manifested.

The Scriptures and the Synagogue

But life was not all persecution, even for the Palestinian Jewish community; and Judaism existed far beyond the bounds of the homeland. Two factors were of immense importance in maintaining the distinctiveness and cohesion of Jews as a religious community and in preserving and applying their inherited religious teaching. On the one hand there were the Scriptures: primarily the five books of Moses, but also the prophetic books and other religious writings. On the other hand there was the Synagogue. Just how old this institution was it is impossible to say; but it is generally held that Synagogues came into existence to meet the needs of the Jews in the Dispersion. Separated from the Temple, the one legitimate centre of sacrificial worship, the exiled Jews could meet at the Synagogue for prayer, for the reading of the Scriptures and instruction in religious truth, and for the transaction of the business of the community.[1] In this there was an important preparation for the time when the Temple was destroyed, and the sacrifices could no longer be offered.

These two factors played a decisive part in the survival of the Jewish faith throughout the centuries. They are also the two most important links between Judaism and Christianity. The early Christian preachers, and the New Testament writers, presented the Gospel as the fulfilment of the Old Testament Scriptures and if this connexion were broken, the Gospel would no longer be the Gospel. On the other hand, the Synagogues scattered across the Graeco-Roman world had disseminated among Gentiles some knowledge of Judaism and provided the Christian preachers with many of their first converts.

Additional Note on the Date of Ezra's Coming to Jerusalem

The following are the chief arguments which have been advanced against the traditional date (458/7 B.C.).

[1] Although it is natural to think of the Synagogue as brought into existence by the Dispersion, in later times at least there were Synagogues in Palestine, and even one within the precincts of the Temple.

1. Ezra came to a settled and populous Jerusalem (Ezra 10¹) protected by a wall (9⁹ A.V., R.V.), whereas Nehemiah had to rebuild the ruined wall (Neh. 1–4) and to draft people into Jerusalem because the population was depleted. But it is not clear that the reference in Ezra 9⁹ is to a city wall in the literal sense: the Revised Standard Version translates the Hebrew word as 'protection'. Further, Ezra's mention of 'a very great assembly' simply means that there was a crowd in a particular place, not that the city was populous.

2. Ezra and Nehemiah are mentioned together in Neh. 8⁹ and 12²⁶, but nowhere else; and there is no reference to Ezra in the passages ascribed to the memoirs of Nehemiah. Arguments from silence are sometimes unconvincing; but if Nehemiah really was a contemporary of Ezra it is remarkable that he does not mention one who was so prominently concerned with the reorganization of community life and the expulsion of foreign elements. Those who press this argument maintain that in Neh. 8⁹ and 12²⁶ Nehemiah's name is an editorial addition.

3. Nehemiah prohibited marriages with foreigners. Ezra insisted that those already contracted should be dissolved. It is argued that Ezra's more stringent policy would follow Nehemiah's rather than precede it. But it is idle to claim that relatively lenient measures (whether they succeed or fail) are always followed by more stringent ones, or that they need have been in this instance.

4. According to the Elephantine papyri, the high priest in Jerusalem in 407 B.C. was called Johanan. In Ezra 10⁶ we read of one Jehohanan or Johanan, who is called the 'son' of Eliashib. The term 'son' is sometimes used rather loosely; and in Neh. 12¹⁰ f., ²² it is stated that Eliashib had a grandson called Jonathan (this name may be a scribal error) or Johanan. From this it is argued that the high priest Johanan of the Elephantine papyri was the contemporary of Ezra and the grandson of Eliashib, Nehemiah's contemporary. This is the strongest argument for rejecting the traditional date of Ezra.

Whether one accepts these arguments as cogent depends in part on whether one thinks that the Chronicler could have made so serious an error in the chronology of a period not very remote from his own (the fourth century?). It must be admitted that his handling of his sources is sometimes confusing. Some scholars hold that the Nehemiah narratives are an addition to the Chronicler's history made by a later editor. On this view, the reversal of the traditional order of Ezra and Nehemiah would present no serious difficulty.

VII

NATURE, MAN, AND GOD

WE saw at the outset that in the Old Testament history and religion are related in a particularly intimate way. The nature of Israel's religion cannot be satisfactorily understood apart from the complex story of Israel's national life and of its relations with other peoples. But in tracing events and developments in their historical sequence it is easy to overlook the importance of some aspects of Israel's faith. We must now briefly consider some of these.

Yahweh was a God who wrought mighty deeds in history. But He was also Lord of nature. This is already evident in the Exodus traditions, in which Yahweh operates through the natural forces as He wills. Later, part of the challenge of the fertility religion of Canaan was that it might have made Israel think of Yahweh simply as a nature god. But Yahweh, though He was indeed giver of the corn, wine, and oil of Canaan, and of the very land itself, was not in any way identified with any or all of the natural processes. He was above them and separate from them. This is clear in the creation stories (Gen. 1^1–2^{4a}, 2^{4b-25}), in the great Psalms of nature and creation (e.g. 8, 19, 104), and in the majestic prophecies of Deutero-Isaiah (e.g. 40^{12-26}). Traces of the mythological content of ancient Near Eastern cosmogonies shine through some of these passages. In Gen. 1 the dividing of the great deep (*tehom*) is in all probability a relic of the myth of the primeval conflict with the chaos monster Tiamat; but, if so, the polytheistic content of the myth has been eliminated, and no one would suppose from the Old Testament text alone that there was any suggestion of a conflict. God is other than nature, Lord of nature, and Creator of nature.

God the Creator is not only the guide of Israel's history, but

Lord of all mankind. This is evident from the universal range
of Gen. 1–11, and also from many passages in the prophets.
There are, however, certain books in the Old Testament which
have a particular international character, and which present
man as man, in his varied relationships as a creature of God.
They are, for the most part, curiously silent about the mighty
deeds of God in Israel's history, and relate their instruction to
His work as Creator and to His providential ordering of the
world. These are the Wisdom books: Proverbs, Job, and Ecclesiastes, and, in the Apocrypha, the Wisdom of Solomon and
Ecclesiasticus. A number of Psalms (e.g. 1, 37, 49, 73) also
belong to the Wisdom literature; and both Joseph and Daniel
reflect the Wisdom ideal.

It has been customary to treat Wisdom teaching as a feature
of post-exilic religion. But although Ecclesiastes is one of the
latest books in the old Testament, and both Job and Proverbs
are post-exilic in their present form, there can be little doubt
that Wisdom was an element in Israelite culture from the
early period of the monarchy onwards. That Job embodies an
ancient tale is generally admitted; and some of the collections
in Proverbs are considerably older than the present form of the
book. There is, indeed, a general probability that the age of
Solomon saw a great flowering of Wisdom lore in Israel. Later
tradition accorded to Solomon the role of Wise Man *par
excellence*. During his reign of peace and outward prosperity,
international communications were good; and this doubtless
facilitated the spread of Wisdom teaching, which was a cosmopolitan phenomenon in the ancient Near East.

Clearly, the sages who imparted Wisdom were a special
class in the community. As the prophet uttered the word of
Yahweh and the priest taught the law (*torah*), so the Wise gave
'counsel' (Jer. 18[18]). In the period of the monarchy, some of
them at least were counsellors at court. The equation of the
Wise with scribes who claim to have the law of Yahweh (Jer.
8[8]) may be linked both with the references to scribes at court
(e.g. 2 Sam. 8[17]; 1 Kings 4[3]; 2 Kings 22[3 ff.]) and also with the
much later passage in Ecclesiasticus which tells of the wisdom

of the scribe who is both counsellor of rulers and student of the law of God (38^{24}–39^{11}).

The Wisdom writings have been called 'the documents of Hebrew humanism'. They delineate the religious and moral responsibility of man as man; and, for the most part, the ideal of character and conduct which they present is that of the wise man, without overt reference to the specifically Israelite tradition. It should, however, be remembered that it was within the life and religious tradition of Israel that these books were written or compiled and handed down. The very use of the divine name Yahweh is a reminder that Israelite Wisdom had national as well as international associations. The later books lay an increasing emphasis on specifically Israelite teaching. In Ecclesiasticus, Wisdom is equated with the *torah* (24).

The multifarious character of Wisdom is reflected in the teaching of the book of Proverbs. For the most part, however, that teaching is directed to the right ordering of human life, with considerable emphasis on the rewards of virtue and prudence and the penalties of various kinds of folly. The various Hebrew terms applied to wisdom and folly, the wise and the fools, indicate a contrast not only between witlessness and astuteness. 'Fools' include the stubborn, self-willed, and arrogant, as well as the stupid and the thoughtless. Clearly, Wisdom was not merely intellectual, but practical and moral. Fundamentally, it was religious. 'The fear of the LORD is the beginning of wisdom' (Prov. 9^{10}; cf. Job 28^{28}). It is the religious element that unifies the whole range of Wisdom teaching: natural lore, rules of conduct, the working of providence, reward and punishment, the ultimate meaning of human life. God's creative Wisdom brought the world into being; and it is by responding to that Wisdom that men can live rightly and fully. This thought is effectively expressed by the vivid personification of Wisdom as God's 'master workman' in creation (Prov. 8; cf. Job 28; Wisd. of Sol. 7^{22}–8^{1}; Ecclus. 24). The metaphorical language used points the way forward to the later conception of a personal divine Agent in the work of creation (cf. John 1^{1-3}, Col. $1^{15\,f.}$).

To questions about reward and punishment and the working of providence Proverbs gives, in general, the answer that wisdom and virtue are rewarded by outward prosperity, whereas folly and vice are punished, a view which has obvious affinities with the Deuteronomic teaching that national religious faithfulness leads to well-being, and apostasy to affliction. But in two Wisdom books this view is subjected to criticism. In Ecclesiastes, orthodoxy alternates with a melancholy scepticism about the purposelessness of life. The interpretation of the book is complicated by the difference of scholarly opinion whether the orthodoxy and the scepticism come from the same author or indicate a drastic process of editorial revision. If the work is substantially by one author, we may perhaps think of him repeating and reflecting on the traditional teaching (e.g. 3^{17}), and at other times denying it in the light of actual experience (e.g. $2^{15\,f.}$), lamenting the seeming purposelessness of life (e.g. $2^{11,\,17}$) and the dismal prospect of what lies beyond it ($3^{20\,f.}$).

In Job the traditional orthodoxy is assailed on a broader front and with a more intense passion. Job, the paragon of piety, is suddenly subjected to material loss, bereavement, and painful disease. Confronted by the conventional view, as expressed by the three friends who come to comfort him, that suffering must be the result of sin, Job obstinately refuses to confess what he has not committed, hurling his challenges and complaints against God, and, at the end, finding peace only when God speaks to him in rebuke out of the whirlwind.

The book gathers together much that is said elsewhere in the Old Testament about the problem of innocent suffering. To reconcile belief in the sovereign righteousness of God with actual experience was a problem made more perplexing by the absence of any effective belief in life after death. Throughout most of the Old Testament the dead are thought to have a shadowy existence in *Sheol*, the counterpart of the Greek *Hades*, but it is mere existence, rather than life (see, e.g. Ps. 6^5, 30^9; Isa. $14^{9\,ff.}$). Whether Job rose above this comfortless expectation is uncertain; for many difficulties beset both the text and the interpretation of the famous passage, 'For I know that my Redeemer lives. . . .' ($19^{25\,ff.}$). But underlying all Job's protests is the longing, not simply for an explanation of his suffering, but for communion with God. When, after the speeches of Job and his friends, God unfolds to Job the wonder of the divine ordering of the universe, the point is not simply that Job is overawed and humbled, but that God has spoken to him, answering his plea, 'Oh, that I knew where I might find him' (23^3). Similarly, the writer of one of the Wisdom Psalms, wrestling with the problem of God's government of the world in the face of the prosperity of the wicked, comes back to his own present communion with God as the effective answer to his perplexity and as outweighing all his affliction (Ps. 73^{23-26}). It seems probable (though some would dispute this) that he sees in this present communion with God a fullness of life which death itself cannot destroy.

The teaching of Israel's sages about God's dealings with man as man, distinctive though it is, is not fundamentally at variance with the central tradition of Israel's religion. That religion is, indeed, intimately bound up with the historical experience of the people Israel; but it is also related to the purpose of God for mankind as a whole. Moreover, in all its main phases it tells of a God with whom men may have personal communion: in the stories of the patriarchs who received His promises and followed His leading; in the experience of Moses to whom He revealed Himself before the deliverance from Egypt; in the call of the prophets who heard His word in the crises of their own times; and in the promise of a time when 'they shall all know Me, from the least of them to the greatest . . .; for I will forgive their iniquity and I will remember their sin no more' (Jer. 31^{34}).

CHRONOLOGICAL TABLES

By R. J. COGGINS

IN the tables which follow, these points should be noted:

1. Precise dating in the history of Israel is impossible before the ninth century. In particular the biblical dates of the kings of Israel and Judah, given in 1 and 2 Kings, contain inconsistencies which have been resolved in various ways. The dates given here are therefore bound to be approximations. For discussion of the main problems, reference must be made to the appropriate sections of each volume.

2. Where names of prophets are given, e.g. Amos, Hosea, this should be understood as referring to the lifetime of the prophet and *not* to the composition of the book which bears his name. This can very rarely be dated with any certainty.

3. The column headed 'Archaeological Evidence' simply lists the main points at which archaeological discovery has thrown light upon the history of Israel. For fuller information, with translations where appropriate, reference should be made to such works as D. W. Thomas (ed.), *Documents from Old Testament Times* (Nelson, 1958), and J. B. Pritchard (ed.), *The Ancient Near East* (O.U.P., 1959).

DATE	ISRAEL	NEIGHBOURING POWERS *Egypt*		ARCHAEOLOGICAL EVIDENCE
1800				
1700	Abraham	Babylonian Power c. 1700	Hyksos Period c. 1720–1550	Mari Documents 1750–1700 Law Code of Hammurabi c. 1700
1600				
1500			XVIIIth Dynasty 1570–1310	
1400	Jacob Descent into Egypt c. 1370	Hurrian (Horite) Power		Nuzu Documents Tablets from Ras Shamra (Ugarit)
1300		Hittite Empire	XIXth dynasty 1310–1200	Tel-el-Amarna Letters
1200	Exodus c. 1250. Moses Entry into Canaan c. 1200 Joshua	✕ Qadesh-Orontes c. 1286	Rameses II 1290–1224 Merneptah 1224–1216 XXth Dynasty 1180–1065 Rameses III 1175–1144 Defeat of the Sea Peoples	Merneptah Stele
1100	Judges Period	Rise of Philistine Power	XXIst Dynasty 1065–935	Wen-Amon c. 1100
1000	Saul c. 1020–1000. Samuel David c. 1000–961 Solomon c. 961–922		XXIInd Dynasty 935–725 Shishak 935–914	Gezer Calendar

PATRIARCHAL PERIOD

	Judah & Prophets	Israel	Egypt	Assyria	Damascus / Babylon	Inscriptions
				Assyria — Revival of Assyrian Power		
	Rehoboam 922–915	Jeroboam I 922–901				
900	Abijam 915–913	Nadab 901–900				
	Asa 913–873	Baasha 900–877			*Damascus*	
		Elah 877–876			Ben-hadad I ?900–860	Melqart Stele
		Zimri 876				
	Jehoshaphat 873–849	Omri 876–869		Shalmaneser III 859–824	Ben-hadad II ?860–843	Black Obelisk of Shalmaneser
	Elijah	Ahab 869–850		✗ Qarqar 853		Moabite Stone
	Jehoram 849–842	Ahaziah 850–849			Hazael 843–796	
800	Ahaziah 842	Joram 849–842				
	Athaliah 842–837	Jehu 842–815				
	Joash 837–800	Elisha	Jehoahaz 815–801		Adad-Nirari III 811–783	
	Amaziah 800–783	Jehoash 801–786			Ben-hadad III ?796–770	
	Azariah (Uzziah) 783–742	Jeroboam II 786–746				
	Amos, Hosea					Samaria Ivories and Ostraca
		Zechariah 746–745				
	Jotham 742–735	Shallum 745		Tiglath-Pileser III 745–727		
	Isaiah (active c. 742–700)	Menahem 745–738			Rezin c. 740–732	
	Ahaz 735–715	Pekahiah 738–737			Fall of Damascus 732	
	Micah	Pekah 737–732				
		Hoshea 732–724		Shalmaneser V 727–722		
				Sargon II 722–705		
		Fall of Samaria 721		Sennacherib 705–681		Siloam Inscription; Taylor Prism of Sennacherib
700	Hezekiah 715–687		XXVth Dynasty 716–663			
	Invasion of Judah 701			Esar-haddon 681–669		
	Manasseh 687–642			Ashur-banipal 669–633?		
			Sack of Thebes 663			
			XXVIth Dynasty 663–525			
			Psammetichus I 663–609		*Babylon*	
	Amon 642–640				Nabopolassar 626–605	
	Josiah 640–609					
	Zephaniah					
	Jeremiah (active 626–c. 580)			Fall of Nineveh 612		Babylonian Chronicle
	Nahum					
	Habakkuk					
	Jehoahaz 609		Necho II 609–593		✗ Carchemish 605	
	✗ Megiddo 609				Nebuchadrezzar II 605–562	
	Jehoiakim 609–598					

DATE	JUDAH	NEIGHBOURING POWERS			ARCHAEOLOGICAL EVIDENCE
		Egypt	Babylon	Persia	
600	Jehoiachin 598–597 (deported) Jerusalem captured 597. First Deportation Zedekiah 597–587[1] Fall of Jerusalem 587[1]; Temple Destroyed; Second Deportation	Psammetichus II 593–588 Apries (Hophra) 588–569			Lachish Letters
	Ezekiel	Amasis 569–525	Amel-Marduk 562–560 Nabonidus 556–539 Fall of Babylon 539		'Jehoiachin' Tablets from Babylon
	'Deutero-Isaiah'			Cyrus 550–530	Cyrus Cylinder
	Third Deportation 581 The Exile				
	Return of some Jews? 537	Egypt conquered by Persia 525		Cambyses 530–522 Darius I 522–486	
	Haggai Zechariah 'Trito-Isaiah'? Temple rebuilt 520–515				
	'Malachi'			Xerxes 486–465 Artaxerxes I 465–424	
	Governorship of Nehemiah 445–433 432– ? Sanballat I, Governor of Samaria	Egypt independent 401		Artaxerxes II 404–358	Elephantine Papyri Samaria Papyri
400	Ezra's Mission 398[2]			Darius III 336–331	
	Alexander the Great conquers Palestine 333–2	Conquests of Alexander: ⨉ Granicus 334; ⨉ Issus 333; ⨉ Gaugamela 331			

[1] = or 586

[2] This may also be dated at either 458 or 428.

NEIGHBOURING POWERS

DATE	JUDAH	Ptolemies	Seleucids	Rome	ARCHAEOLOGICAL EVIDENCE
300	Ptolemies rule Palestine	Ptolemy I Soter 323–285 Ptolemy II Philadelphus 285–246 Ptolemy III Euergetes 246–221 Ptolemy IV Philopator 221–203	Seleucus I 312–281 Antiochus I 281–261 Antiochus II 261–247 Seleucus II 247–226 Seleucus III 226–223 Antiochus III 223–187 ✕ Raphia 217		Zeno Papyri
200	Seleucids rule Palestine Profanation of the Temple 167 (?168) Maccabaean Revolt Book of Daniel 167/4 Rededication of the Temple 164 (?165)	Ptolemy V Epiphanes 203–181	✕ Panium 198 Seleucus IV 187–175 Antiochus IV Epiphanes 175–163		
100	Hasmonean Rulers Judas Maccabaeus 166–160 Jonathan 160–143 Simon 142–134 John Hyrcanus I 134–104 Aristobulus I 104–103 Alexander Jannaeus 103–76 Alexandra Salome 76–67 Aristobulus II 67–63 Pompey captures Jerusalem 63 Judah added to the Roman Province of Syria Hyrcanus II 63–40 Antigonus 40–37 Herod the Great 37–4 B.C.		Antiochus V Eupator 163–162 Demetrius I Soter 162–150 Alexander Balas 150–145 Demetrius II Nicator 145–139, 129–125 Antiochus VY Epiphanes 145–142 (Tryphon 142–139) Antiochus VII Sidetes 139–128	?Qumran Sect established Jewish Independence granted 142 Overthrow of Pompey 48 ✕ Philippi 42 ✕ Actium 31	Qumran Scrolls (?)

INDEX OF SCRIPTURE REFERENCES

SUBJECT INDEX